THE FACTS ON FILE CHEMISTRY HANDBOOK

THE FACTS ON FILE CHEMISTRY HANDBOOK

THE DIAGRAM GROUP

Checkmark Books®
An imprint of Facts On File, Inc.

The Facts On File Chemistry Handbook

Diagram Visual Information Ltd

Editorial director	Moira Johnston
Editors	Nancy Bailey, Jean Brady, Paul Copperwaite, Eve Daintith, Bridget Giles, Jane Johnson, Reet Nelis, Jamie Stokes
Design	Richard Hummerstone, Edward Kinsey
Design production	Carole Dease, Oscar Lobban, Lee Lawrence
Artists	Susan Kinsey, Lee Lawrence, Kathleen McDougal
Research	Peter Dease, Catherine & Neil McKenna,
Contributors	Michael Allaby, Martyn Bramwell, John Daintith, Trevor Day, John Haywood, Jim Henderson, David Lamber Catherine Riches, Dr Robert Youngson
Indexer	Christine Ivamy

Checkmark Books
An imprint of Facts On File, Inc.
11 Penn Plaza
New York NY 10001

Library of Congress Cataloging-in-Publication Data

The Facts on File chemistry handbook / The Diagram Group.
p. cm.
Includes index
ISBN 0-8160-4080-X (hc) (acid-free paper)—ISBN 0-8160-4585-2 (pbk)
I. Chemistry—Handbooks, manuals, etc. I. Diagram Group.

QD65 .F33 2000
540—dc21

99-048563

Checkmark Books are available at special discounts when purchased in bulk quantities for businesses, associations, institutions, or sales promotions. Please call our Special Sales Department in New York at 212/967-8800 or 800/322-875

You can find Facts On File on the World Wide Web at http://www.factsonfile.co

Cover design by Cathy Rincon

Printed in the United States of America

MP DIAG 10 9 8 7 6 5 4 3 2
(pbk) 10 9 8 7 6 5 4 3 2 1

This book is printed on acid-free paper.

INTRODUCTION

An understanding of science is the basis of all technological advances. Our domestic lives, possessions, cities, and industries have only been developed through scientific research into the principles that underpin the physical world. But obtaining a full view of any branch of science may be difficult without resorting to a range of books. Dictionaries of terms, encyclopedias of facts, biographical dictionaries, chronologies of scientific events – all these collections of facts usually encompass a range of science subjects. THE FACTS ON FILE HANDBOOK LIBRARY has coverage of four major scientific areas – CHEMISTRY, PHYSICS, EARTH SCIENCE (including astronomy), and BIOLOGY.

THE FACTS ON FILE CHEMISTRY HANDBOOK contains four sections – a glossary of terms, biographies of personalities, a chronology of events, essential charts and tables, and finally an index.

GLOSSARY
The specialized words used in any science subject mean that students need a glossary in order to understand the processes involved. THE FACTS ON FILE CHEMISTRY HANDBOOK glossary contains more than 1,400 entries, often accompanied by labeled diagrams to help clarify the meanings.

BIOGRAPHIES
The giants of science – Darwin, Galileo, Einstein, Marie Curie – are widely known, but hundreds of other dedicated scientists have also contributed to scientific knowledge. THE FACTS ON FILE CHEMISTRY HANDBOOK contains biographies of more than 300 people, many of whose achievements may have gone unnoticed but whose discoveries have pushed forward the world's understanding of chemistry.

CHRONOLOGY
Scientific discoveries often have no immediate impact. Nevertheless, their effects can influence our lives more than wars, political changes, and world rulers. THE FACTS ON FILE CHEMISTRY HANDBOOK covers nearly 9,000 years of events in the history of discoveries in chemistry.

CHARTS & TABLES
Basic information on any subject can be hard to find, and books tend to be descriptive. THE FACTS ON FILE CHEMISTRY HANDBOOK puts together key charts and tables for easy reference. Scientific discoveries mean that any compilation of facts can never be comprehensive. Nevertheless, this assembly of current information on chemistry offers an important resource for today's students.

In past centuries scientists were curious about a wide range of sciences. Today, with disciplines so independent, students of one subject rarely learn much about others. THE FACTS ON FILE HANDBOOKS enable students to compare knowledge in chemistry, physics, earth science, and biology, to put each subject in context, and to underline the close connections between all the sciences.

CONTENTS

SECTION ONE GLOSSARY

A(r) Symbol for relative atomic mass.

absolute temperature (thermodynamic temperature) Based on absolute zero. The unit (the kelvin) is 1/273.16 of the temperature of the triple point of water and is equivalent to one degree Celsius (1°C).

absolute zero The lowest possible temperature. Zero on the Kelvin scale.

abundance A measure of the quantity of a substance occurring in a particular area (an element in the Earth's crust or an isotope in a sample of an element). It is expressed in percentage or parts per million.

Ac Symbol for the element actinium.

accelerator A chemical that increases the rate of a chemical reaction.

accumulator or **battery** A device that uses chemical energy to store electrical energy.

acetaldehyde *See* ethanal.

acetic acid *See* ethanoic acid.

acetone *See* propanone.

acetylene *See* ethyne.

acid Any substance that releases hydrogen ions when added to water. It has a pH of less than 7.

acid anhydrides Compounds that react with water, forming acids, for example, the acid anhydride SO_2 that reacts to make the acid H_2SO_4.

acid-base reaction An acid and a base react together to form a salt and water only.

acidic oxide The oxides of nonmetals that form acidic solutions in water. An acidic oxide reacts with a base to form salt and water only.

acidification The fall in pH in a solution caused by the addition of an acid. This is seen in nature in the pollution of lakes, rivers, and groundwater by acid rain.

acid—organic *See* organic acid.

acid rain A form of pollution where rain dissolves acidic gases (mainly sulfur dioxide) from the air. Sulfur dioxide is released into the atmosphere by the burning of fossil fuels.

acid salt A salt of a polybasic acid in which not all the hydrogen atoms have been replaced by a metal or metal-like group (such as the ammonium group).

acid—standardization of *See* standardization of solutions.

actinides (actinoids) The name of the group of elements with atomic numbers from 89 (actinium) to 103 (lawrencium). All are radioactive and have similar properties to actinium. As their outer electronic structure is very similar (the f orbital in their fifth shell is being filled), they have similar chemical properties.

actinium Element symbol, Ac; silvery metallic element; Z 89; A(r) 227; density (at 20°C), 10.07 g/cm^3; m.p., 1050°C; radioactive; name derived from the Greek *aktis*, "ray"; discovered 1899.

actinium series One of the naturally occurring radioactive series.

activated complex A short-lived association of atoms that is formed during a chemical reaction.

activation energy The energy barrier to be overcome in order for a reaction to occur. Many chemical reactions require heat energy to be applied to reactants to initiate a reaction.

active carbon Particles of carbon used widely as an adsorbent to remove impurities in gases and liquids.

addition polymerization A process by which molecules join together by a series of addition reactions to form larger molecules, or macromolecules, which consist of repeated structural units.

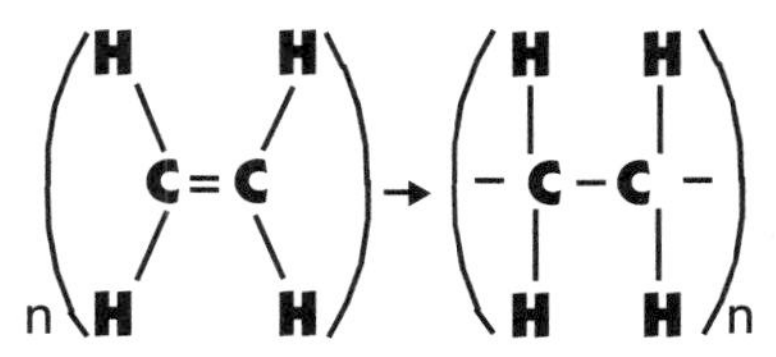

Addition polymerization

addition reaction A reaction in which a molecule of a substance reacts with another molecule to form a single compound. The term addition reaction is often used in organic chemistry to describe a reaction in which an atom is added to either side of the double or triple bond in an unsaturated compound to form a saturated compound.

additive A small quantity of a compound added to a bulk material to give it certain properties. For example, the colorings added to food and drink.

adsorption The process by which molecules of gases or liquids become attached to the surface of another substance. Desorption is the opposite process.

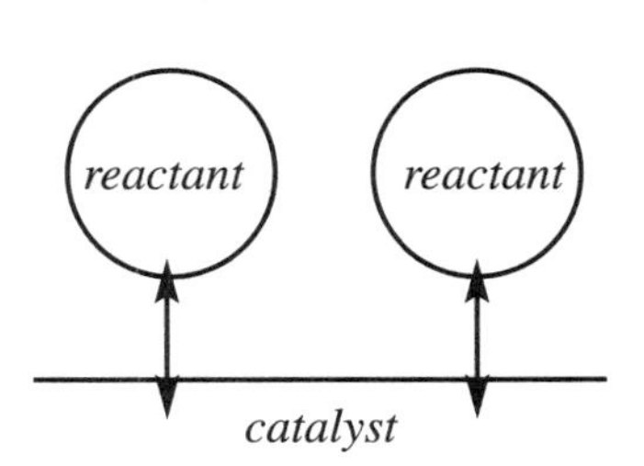

Adsorption

aerosol Extremely small liquid or solid particles suspended in air or another gas.

Ag Symbol for the element silver.

agrochemicals Chemicals used in agriculture, with the exception of fertilizers. The classification includes fungicides, herbicides, pesticides, growth regulators, and vitamin and mineral supplements.

air—a mixture Air is a mixture of several gases (*see* air—composition of). These can be physically separated by cooling (to remove water vapor) and by fractional distillation (to remove nitrogen). The properties of air are an average of its components.

air—composition of The composition of air varies but its average composition (given in percentages by volume) is nitrogen, 78; oxygen, 21; argon, 0.93; carbon dioxide, 0.03.

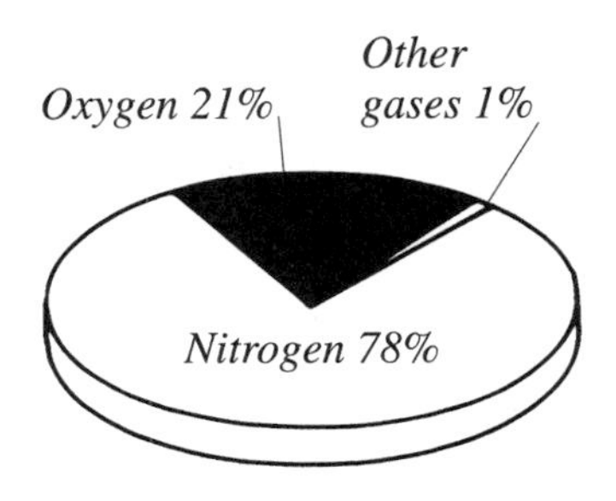

Composition of air

air—liquid Liquid air is a pale blue liquid that boils at –193°C. As its component parts have different boiling points (nitrogen boils at –195.8°C, oxygen boils at –183°C), nitrogen and oxygen can be obtained by the fractional distillation of liquid air.

Al Symbol for the element aluminum.

alcohols *See* alkanols.

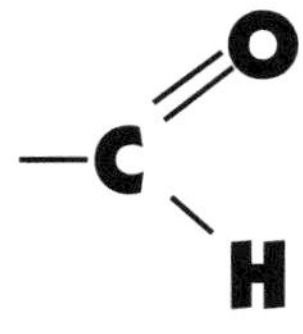

Aldehyde

aldehydes A group of organic compounds containing the aldehyde group (–CHO). Names have the suffix -al.

algae A loose grouping of plant-like organisms including many single-celled forms and multicellular forms such as seaweeds.

algal bloom A rapidly growing layer of algae that floats on the surface of a body of water and whose growth is stimulated by nitrates and phosphates in fertilizers. This layer can cause plants growing at the bottom of the water to die as the light they need is shielded from them by the algal bloom.

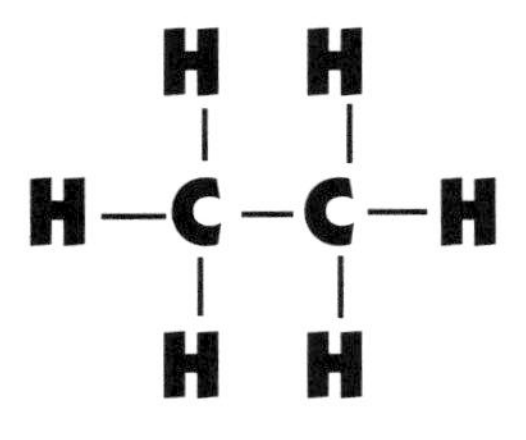

Alkane (ethane)

aliphatic Organic compounds composed of carbon atoms in straight or branched chains.

alkali A solution of a substance in water that has a pH of more than 7 and has an excess of hydroxide ions in the solution.

alkali metals Metallic elements found in group 1 of the periodic table. They are very reactive, electropositive, and react with water to form alkaline solutions.

alkaline earth metals Metallic elements found in group 2 of the periodic table. They are less reactive and electropositive than alkali metals but also produce alkaline solutions when they react with water.

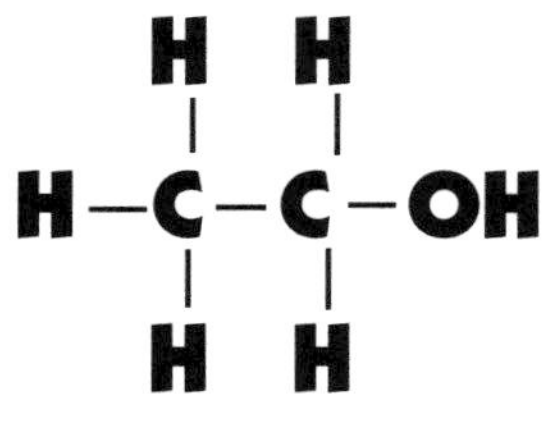

Alkanol (ethanol)

alkali—standardization of *See* standardization of solutions.

alkanal An aldehyde in which the radical attached to the aldehyde group is aliphatic.

alkanes A group of hydrocarbons whose general formula is C_nH_{2n+2}. They have single bonds between the carbon atoms and are thus said to be saturated and hence not very reactive.

alkanols (alcohols) A family of organic compounds whose structure contains the –OH functional group. General formula $C_nH_{2n+1}OH$.

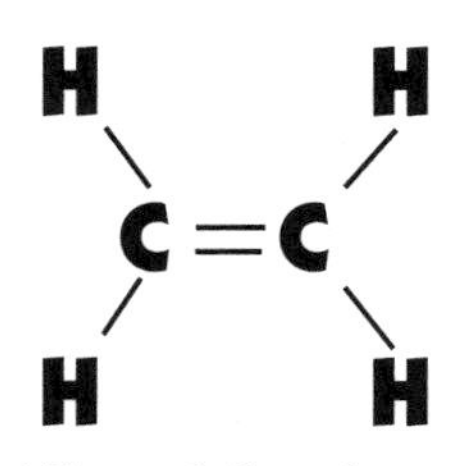

Alkene (ethene)

alkenes A group of hydrocarbons, the general formula of which is C_nH_{2n}. They are unsaturated, having a double bond between a pair of carbon atoms. They are reactive and undergo addition reactions.

alkene—substituted An alkene in which one or more hydrogen atom has been replaced by a different atom (or group of atoms).

alkylation The insertion of alkyl groups into either hydrocarbon chains or aromatic rings.

alkyl group A hydrocarbon group whose general formula is C_nH_{2n+1}.

alkynes A group of hydrocarbons whose general formula is C_nH_{2n-1}. They are unsaturated, having a triple bond between a pair of carbon atoms in each molecule and are thus reactive, undergoing addition reactions.

allo- A prefix to the name of a chemical compound that shows that the compound is a stereoisomer of a more common compound.

allotrope An element that can exist in more than one physical form while in the same state. Carbon can occur in two common allotropes, diamond and graphite (a third – buckminsterfullerene – has been discovered recently). The physical properties of these allotropes are very different.

alloy A metallic material made of two or more metals or of a metal and nonmetal. By mixing metals in certain proportions, alloys with specific properties can be made.

alpha particle A particle released during radioactive decay. It consists of two neutrons and two protons (the equivalent of the helium atom). Energy is released by this change; most is accounted for by the kinetic energy of the a particle that moves away at high speed but that rapidly loses energy by collision and ionization of other atoms and molecules and is easily stopped by a piece of paper. Alpha rays are streams of fast-moving a particles.

alumina A naturally occurring form of aluminum oxide also known as corundum.

aluminum Element symbol, Al; group 3; silvery white metallic element; Z 13; A(r) 26.98; density (at 20°C), 2.70 g/cm^3; m.p., 660.4°C; name derived from the Latin *alumen*; discovered 1827.

aluminum chloride $AlCl_3$. Anhydrous aluminum chloride fumes in moist air, reacting to form hydrogen chloride with water vapor.

aluminum hydroxide $Al(OH)_3$. A white crystalline compound. It appears as a white or yellowish gelatinous mass on precipitation from solutions of ammonium salts, in which form it contains coordinated water molecules and water molecules trapped in its structure. Partially dried gels of aluminum hydroxide are used as drying agents, catalysts, and absorbents.

aluminum nitride AlN. Formed (together with the oxide) when aluminum is heated strongly.

aluminum oxide Al_2O_3. A white or colorless crystalline compound. It is formed by heating aluminum hydroxide and has two main forms, the alpha form and the gamma form. The alpha form occurs naturally and is known as corundum. The gamma form (activated alumina) is used as a catalyst as it has adsorptive properties. Bauxite is a hydrated form of aluminum oxide. Aluminum oxide is amphoteric. It reacts with sodium hydroxide to form sodium aluminate ($NaAlO_2$) and water, and with hydrochloric acid to form aluminum chloride and water.

Amides (ethanamide)

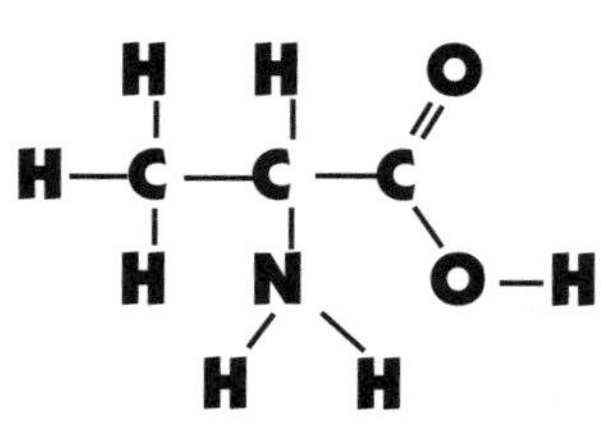

Amino acid

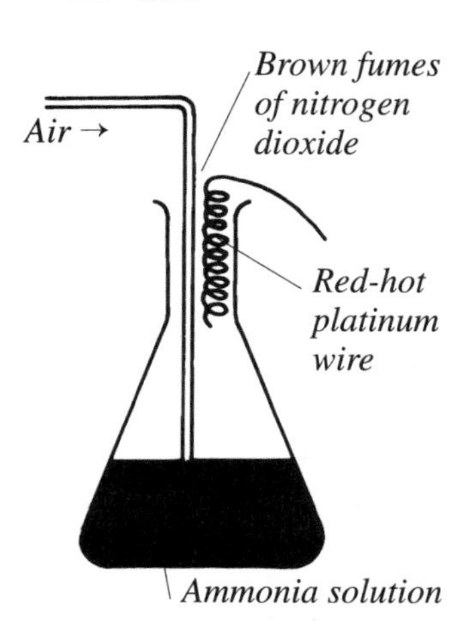

Oxidation of ammonia

Am Symbol for the element americium.

amalgam An alloy containing mercury.

americium Element symbol, Am; Actinide; silvery radioactive metallic element; Z 95; A(r) 243; density (at 20°C), 13.67 g/cm³; m.p., 994°C; name named for America; discovered 1944.

amide group $CONH_2$. A functional group consisting of a carbon atom joined to an oxygen atom with a double bond and to a nitrogen atom that is joined to two hydrogen atoms.

amides A group of organic compounds containing the amide group. Their general formula is $RCONH_2$. Amides are white solids that are soluble in alcohol and ether; some are soluble in water.

amines A group of organic compounds containing the amino functional group $-NH_2$.

amino acids A group of organic compounds containing both the carboxyl group (–COOH) and the amino group ($-NH_2$).

amino group $-NH_2$.

ammonia NH_3. Colorless, strong-smelling poisonous gas, very soluble in water, forming a weak alkaline solution. $NH_3 + H_2O = NH_4^+ + OH^-$. It burns in oxygen with a yellowish flame. It is used industrially in the manufacture of fertilizers and the production of nitric acid. Most ammonia used is produced by the Haber process.

ammonia—eighty-eight A concentrated solution of ammonia in water that contains about 35% by mass of ammonia. Its relative density is 0.880 – hence the name. It softens water and helps to remove stains from clothes.

ammonia—liquor A solution of ammonia in water that is produced during coal-gas manufacture. It is used to make the fertilizer ammonium sulfate.

ammonia—oxidation If air is passed through a solution of ammonia in a flask and a red-hot platinum wire is placed at the top of the flask, the ammonia reacts with the oxygen in the air to form nitrogen monoxide. This then reacts with more oxygen to form brown fumes of nitrogen dioxide. As the reaction is exothermic, the platinum wire continues to glow red during the reaction.

ammonia—reactions Ammonia is very soluble in water, forming a weak alkaline solution. $NH_3 + H_2O = NH_4^+ + OH^-$. Ammonia burns in oxygen with a yellowish flame and reacts with acids to form ammonium salts. Ammonium salts contain the ammonium ion NH_4^+.

ammonia—soda process *See* Solvay process.

ammonia—solution Ammonia solution is a weak alkali. It precipitates insoluble hydroxides from metal salts in solution. (*See* ammonium hydroxide.)

ammonium carbonate $(NH_4)_2CO_3$. Formed as a sublimate (mixed with ammonium hydrogen carbonate) when calcium carbonate and ammonium sulfate (or chloride) are heated together. It is very soluble in water. Ammonium carbonate decomposes to form NH_3, CO_2, and H_2O on heating and decomposes in moist air to form ammonium hydrogen carbonate. It smells of ammonia, and the mixture of ammonium carbonate and ammonium hydrogen carbonate is also called sal volatile. The mixture is used in smelling salts and baking powder.

ammonium chloride NH_4Cl (also called sal ammoniac) A white crystalline solid that is soluble in water. It sublimes on heating to form ammonia and hydrogen chloride (gas). It is used in dry cells, as a flux in soldering, and as a mordant.

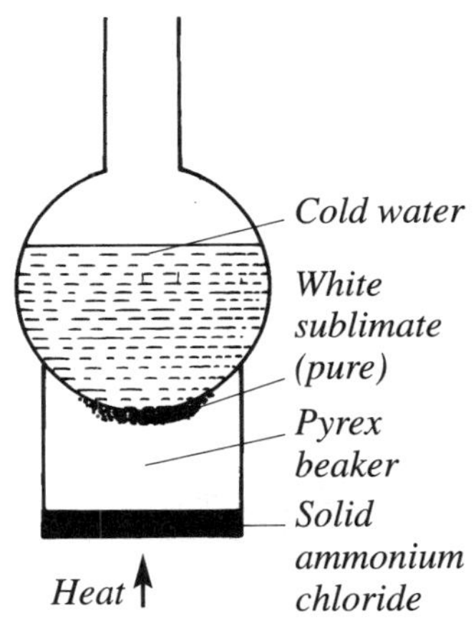

Sublimation of ammonium chloride

ammonium hydrogen carbonate NH_4HCO_3. A white crystalline solid. It is more stable than ammonium carbonate and is therefore often used in its place both medicinally (smelling salts) and in baking powders.

ammonium hydroxide NH_4OH. It exists as an aqueous solution of ammonia and it contains ammonium ions, hydroxide ions, unionized ammonia, and water.

ammonium ion NH_4^+. Found in ammonia solution and in ammonium compounds. Ammonium salts are similar to the salts of monovalent metals.

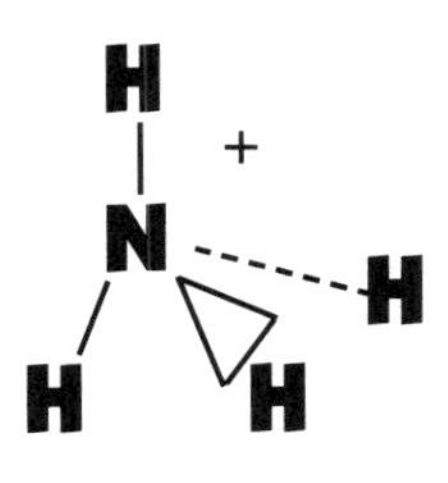

Ammonium ion

ammonium nitrate NH_4NO_3. A colorless crystalline solid that dissolves readily in water. This is an endothermic reaction (the solution becomes cold), and a mixture of ammonium nitrate and water can be used as a freezing mixture (*see* freezing). When heated, ammonium nitrate forms dinitrogen oxide. Ammonium nitrate is used as a fertilizer and also as an explosive, with a suitable detonator, although it can detonate spontaneously.

ammonium nitrite NH_4NO_2. Very unstable; decomposes to form nitrogen and water.

ammonium salt—test for Into a test tube containing a small amount of an aqueous solution of a base, carefully add a small amount of the compound to be tested. Add more of the compound if there is no reaction. If the compound dissolves in cold alkali and liberates a gas that turns red litmus paper blue, this indicates that the gas is ammonia and that the compound tested is an ammonium salt. $NH_4^+ + OH^- = NH_3 + H_2O$.

ammonium sulfate $(NH_4)_2SO_4$. A colorless crystalline solid. It has been used as fertilizer but is now being replaced by fertilizers with higher nitrogen content. It is produced by passing ammonia and carbon dioxide into a suspension of calcium sulfate (gypsum).

amorphous Lacking form, shape, or crystal structure: amorphous substances have no fixed melting point.

amphoteric Exhibiting properties of both an acid and a base. An amphoteric compound reacts with both acids and bases to form salts.

analysis A method of finding out what the component parts of a material are. *See* qualitative analysis and quantitative analysis.

anaerobic A process that takes place in the absence of free oxygen.

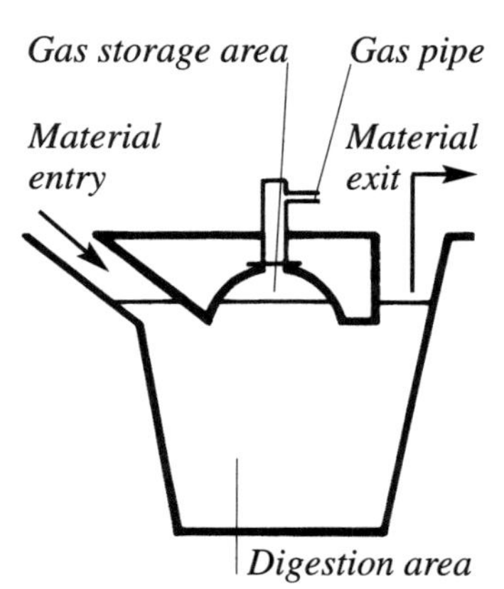

Anaerobic production of biogas

anesthetic A substance used to relieve pain. General anesthetics affect the whole body, producing unconsciousness. Local anesthetics affect a specific part of the body.

anhydride The substance remaining when one or more molecules of water have been removed from an acid (or a base). Most anhydrides are good drying agents.

anhydrite Calcium sulfate ($CaSO_4$), which occurs naturally in an anhydrous state.

anhydrous Containing no water. Term applied to salts without water of crystallization.

anion An ion having a negative charge.

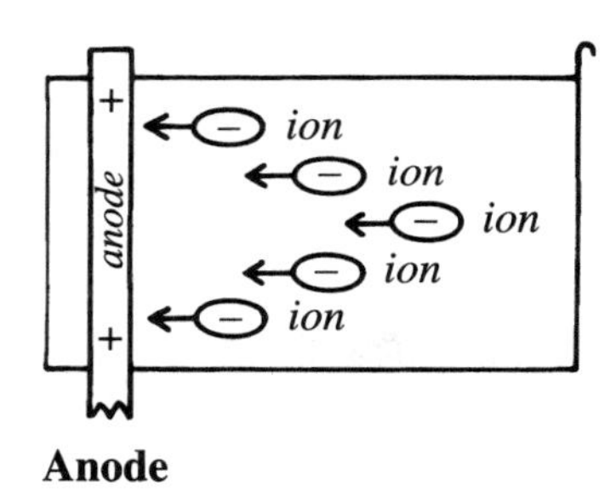

Anode

annealing A method of treating materials (metals and glass) to increase their strength and to relieve strain in their structure. The material is heated to a high temperature and then cooled slowly. In metals, this process causes large crystals to form, increasing the metal's malleability.

anode The electrode carrying the positive charge in a solution undergoing electrolysis.

anodize To coat the surface of a metal with a film of protective oxide. This can be done by making the metal the anode in an electrolysis cell.

antacid A substance such as milk of magnesia (MgO) and sodium bicarbonate ($NaHCO_3$) that is taken to neutralize excessive stomach acid in order to relieve indigestion.

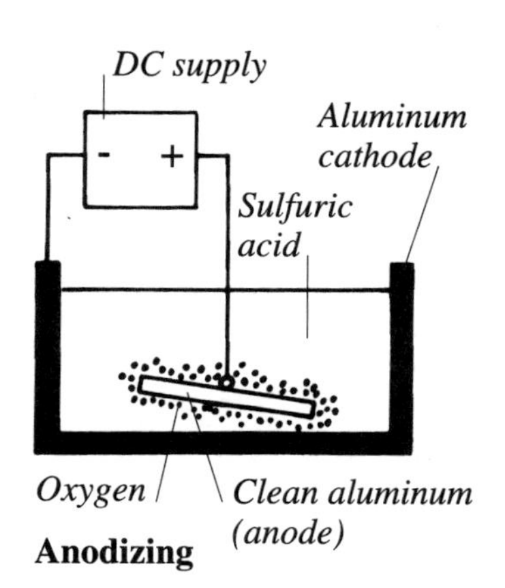

Anodizing

anti-foaming powder A substance that is used in a number of processes to reduce the quantity of foam produced in order to increase the efficiency of the process. Polyamides are used in boiler systems. Low concentrations of silicones are also used widely.

antifreeze A substance that lowers the freezing point of water. Methanol and ethane-1,2-diol are examples of antifreeze agents that are added to the cooling systems of engines to prevent damage that would be caused by the formation of ice. A concentration of 30% methanol and water or 35% ethane-1,2-diol and water will remain liquid above –20.6°C.

antiknock In an internal combustion engine, gasoline and air must explode together at the correct moment or preignition occurs, making "knocking" sounds as the fuel explodes prematurely. Antiknock agents are added to the fuel to overcome this problem. They promote more efficient combustion (and increase the octane rating of the fuel). Lead(IV) tetraethyl has been an important antiknock agent, but it is being withdrawn because of problems with lead pollution. *See* octane rating.

antimony Element symbol, Sb; group 5; most stable form has bluish white metallic appearance; derivatives very toxic; Z 51; A(r) 121.75; density (at 20°C), 6.68 g/cm^3; m.p., 630.7°C; Latin name, *antimonium-aktis*, "ray"; discovered before 1600; antimony compounds used in flame proofing, ceramics, and dyestuffs.

anti-oxidants Chemical additives that slow down the rate at which a substance is degraded by oxidation. When used in food, they increase the length of time a product can be kept. They are also added to paint and plastics.

aqua fortis Concentrated nitric acid.

aqua regia A mixture of one part concentrated nitric acid and three parts of concentrated hydrochloric acid. It dissolves all metals except silver.

aqueous solution A solution in which water is the solvent.

Ar Symbol for the element argon.

arene The general name for an aromatic hydrocarbon.

argon Element symbol, Ar; noble gas, group 8; Z 18; A(r) 39.95; density (at 20°C), 1.784 g/l at STP; m.p., –189.2°C; name derived from the Greek *argos*, "inactive"; discovered 1894; used in light bulbs.

aromatic compounds The group of hydrocarbons derived from benzene (C_6H_6), that have a ring structure.

arsenic Element symbol, As; group 5; a metalloid with bright metallic

Aromatic compound

appearance; Z 33; A(r) 74.92; density (at 20°C), 5.73 g/cm^3; m.p., 817°C; arsenic compounds poisonous; name derived from the Greek *arsenikon*; discovered 1250; used in insecticides, semiconductors, and in alloys where it has a hardening effect.

aryl An aromatic hydrocarbon group formed by the removal of a hydrogen atom from an arene.

As Symbol for the element arsenic.

asbestos A naturally occurring fibrous material consisting mainly of calcium magnesium silicate. It has heat- (insulating and fire) proofing properties and was formerly widely used. It is now known to cause both asbestosis (a lung disease) and mesothelioma (a tumor of the epithelium lining the lungs, abdomen, or heart associated with exposure to asbestos) and its use has been greatly restricted.

association The process by which molecules of a substance combine to form a larger structure. This occurs in liquid ammonia where the liquid consists of $(NH_3)x$ molecules rather than separate NH_3 molecules. An associated liquid is formed when molecules of one substance are held together with molecules of another by forces weaker than normal chemical bonds. For example, a mixture of ethanol and water forms an associated liquid in which the molecules are held together by hydrogen bonds.

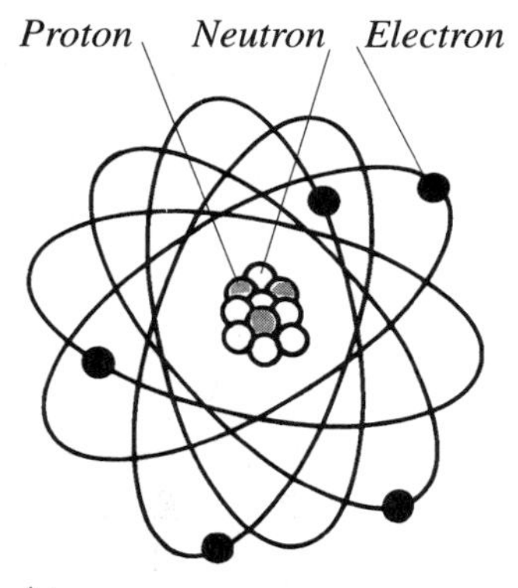

Atom

astatine Element symbol, At; group 7, halogen; radioactive; Z 85; A(r) 210; m.p., 302°C; name derived from the Greek *astatos*, "unstable"; discovered 1940.

At Symbol for the element astatine.

atom The smallest particle of an element that can exhibit that element's properties. An atom has a small, massive nucleus of protons and neutrons surrounded by a cloud of electrons (equal in number to the number of protons in the nucleus and unique to the element).

atomic energy The energy liberated by changes in the nuclei of atoms. When the nuclei of radioactive elements break up and other elements are formed, matter is destroyed. This matter is converted to energy in the formula $E = mc^2$. (One kilogram of matter yields 9×10^{16} joules of energy.)

atomicity The atomicity of an element is the number of atoms in one molecule of the element. For oxygen (O_2) it is 2; for ozone (O_3) 3; for hydrogen (H_2) 2.

atomic mass Short for relative atomic mass.

atomic mass unit Defined as 1/12 the mass of one atom of carbon-12 isotope.

atomic number or **proton number** (Z) The number of protons in the nucleus of an atom. If not electrically charged, this is equal to the number of electrons in its shells.

atomic orbital *See* orbital.

atomic theory Matter consists of atoms, which are made of electrons, protons, and neutrons. Atoms can be created and destroyed in radioactive changes but not in chemical reactions. All atoms of an element contain the same number of protons. Atoms of an element may differ in mass because they contain different numbers of neutrons (*see* isotope). These do not affect their chemical properties. Chemical combination usually occurs between small, whole numbers of atoms (although it can occur between very large numbers of atoms, particularly with carbon compounds – *see* polymerization).

Au Symbol for the element gold.

Aufbau principle This governs the order in which orbitals are filled in successive elements in the periodic table: 1s, 2s, 2p, 3s, 3p, 4s, 3d, 4p, 5s, etc. The number is the shell number and the letter denotes the orbital type.

autocatalysis The action as a catalyst by one of the products of a chemical reaction.

autoclave A strong vessel in which substances may be heated under pressure in order to carry out reactions at high temperatures and pressures. Autoclaves are also used for sterilization of equipment.

Avogadro constant or **number** (L) The number of particles (atoms, molecules, ions) present in a mole of substance. Specifically, it is the number of atoms present in 12 g of the carbon-12 isotope (6.023×10^{23}).

Avogadro's hypothesis or **law** Equal volumes of all gases at the same temperature and pressure contain the same number of molecules.

azeotrope (azeotropic mixtures) A mixture of liquids that boils without a change in composition, i.e. when it boils it gives off a vapor whose composition is the same as the liquid.

azides Compounds that contain the ion N_3^- or the group $-N_3$. Heavy metal azides are explosive.

azo compound A compound that contains two aromatic rings connected by an azo group. Many azo compounds are dyes.

azo group -N=N- An organic group containing two nitrogen atoms.

B Symbol for the element boron.

Ba Symbol for the element barium.

bakelite A phenol/methanal resin that was patented in 1909 by Leo Hendrik Baekeland. Bakelite is dark in color and has good electrical and heat

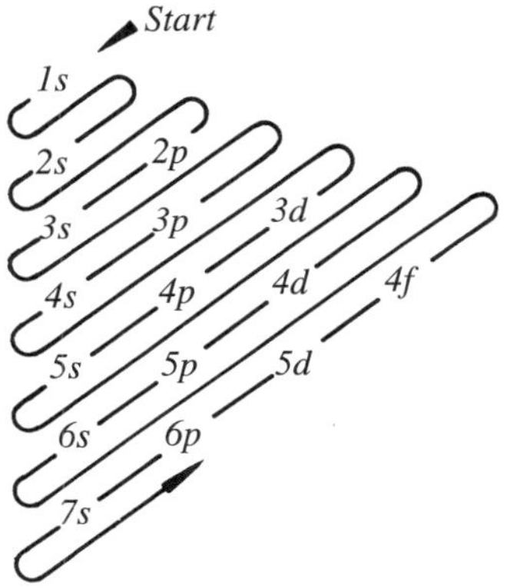

Aufbau principle

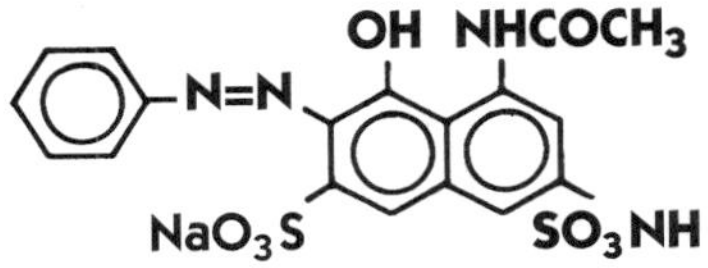

Azo compound

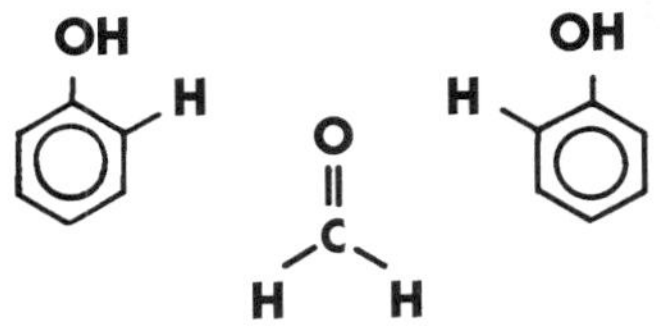

Bakelite

insulation properties. It has been used as a covering for electric plugs and switches, for the handles of saucepans and other household items, jewelry, and more.

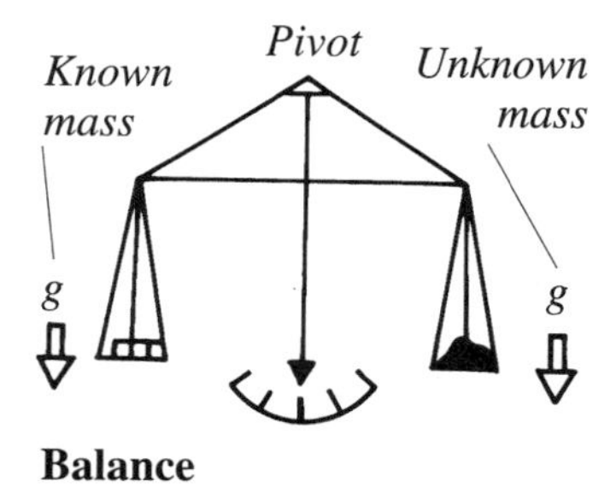

Balance

baking powder A mixture that produces carbon dioxide when heated or wetted. It is usually a mixture of sodium hydrogencarbonate and tartaric acid, or cream of tartar. If baking powder is mixed with other ingredients, the carbon dioxide produced causes the mixture to rise.

baking soda Sodium hydrogencarbonate. When heated, it decomposes to form sodium carbonate, carbon dioxide and water. $2NaHCO_3 = Na_2CO_3 + CO_2 + H_2O$. If baking soda is mixed with other ingredients, the carbon dioxide produced causes the mixture to rise on cooking.

balance An instrument for comparing the masses of objects.

balanced equation A balanced chemical equation has equal numbers of each atom on each side of the equation. Such an equation can be used to calculate the masses of substances either reacting or being produced in a chemical reaction. To do this, it is assumed that each formula represents one mole of the substance, and weights can then be substituted where known to calculate the unknown quantities.

barite or **barytes** The mineral form of barium sulfate, a useful source of barium compounds.

barium Element symbol, Ba; group 2, alkaline earth metal; silver white metal; Z 56; A(r) 137.33; density (at 20°C), 3.5 g/cm^3; m.p., 725°C; compounds poisonous and opaque to X rays; name derived from the Greek *barys*, "heavy"; discovered 1808; used as a getter to remove oxygen, salts used in X-ray diagnosis.

barium carbonate $BaCO_3$. A white insoluble compound that occurs in the mineral witherite. It is used to make other barium salts, flux for ceramics, and in the manufacture of some types of optical glass.

barium chloride $BaCl_2$. A poisonous white compound that is used for the electrolytic production of barium.

barium chromate $BaCrO_4$. A yellow pigment that is fairly insoluble in water.

barium hydroxide $Ba(OH)_2$ (baryta) A white solid that is sparingly soluble in water. It is used in the laboratory as a weak alkali in volumetric analysis. It is also used as a plastic stabilizer and a gasoline additive.

barium peroxide BaO_2. A dense off-white solid that is used as a bleaching agent and in the manufacture of hydrogen peroxide.

barium sulfate $BaSO_4$. A poisonous white solid that is insoluble in water. Its mineral form is barytes. It is used as a pigment and as an additive in the glass and rubber industries. It is administered orally (barium sulfate is safe to use as it is very insoluble) for X-ray investigations.

base (usually a metal oxide or hydroxide) A substance existing as molecules or ions that can take up hydrogen ions. When a base reacts with an acid it forms a salt and water only.

base—equivalent of The mass in grams that reacts with the equivalent weight of an acid (1.08 g of hydrogen ions).

base—standardization of *See* standardization of solutions.

basic Having the properties of a base.

basicity of acids The number of hydrogen ions formed by a molecule of an acid. Hydrochloric acid (HCl) is monobasic. Sulfuric acid (H_2SO_4) is dibasic. Phosphoric acid (H_3PO_4) is tribasic.

basic oxide Many metal oxides are basic. Basic oxides react with acids, forming a salt and water only.

basic oxygen furnace A vessel in which a blast of oxygen is passed over the surface of, or through, molten iron to convert it to steel.

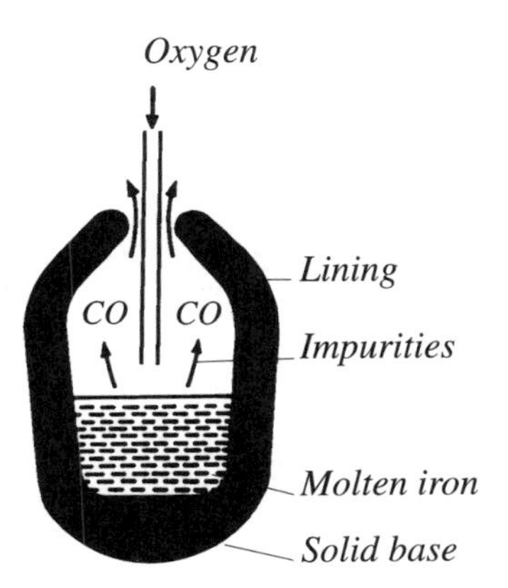

Basic oxygen furnace

battery *See* accumulator.

bauxite The ore from which aluminum is extracted. It is a hydrated form of aluminum oxide ($Al_2O_3 . \times H_2O$).

Be Symbol for the element beryllium.

Benedict's solution A blue solution used to test for reducing sugars. It contains copper(II) sulfate, sodium, carbonate, and sodium citrate.

Benedict's test If a mixture of an aqueous solution of a reducing sugar and Benedict's solution is heated, when the temperature approaches boiling point, the color of the mixture changes from blue to green-yellow or orange. A brick red precipitate of copper(I) oxide is then formed.

Benzene (abbreviated form)

benzene An aromatic hydrocarbon produced from naphtha. Its formula is C_6H_6, and each of the six hydrogen atoms is attached to one of the six carbon atoms that are arranged at the corners of a hexagon. This arrangement is called a benzene ring. Benzene is an important source of other organic compounds.

berkelium Element symbol, Bk; actinide; Z 97; A(r) 247; density (at 20°C), 14 (est) g/cm^3; m.p., 986°C; named for Berkeley, California; discovered 1950.

beryllium Element symbol, Be; alkaline earth metal, group 2; gray, hard brittle metal; Z 4; A(r) 9.01; density (at 20°C), 1.85 g/cm^3; m.p., 1287°C; compounds toxic; name derived from the Greek *beryllos*, "beryl"; discovered 1798; used in alloys and in nuclear reactors.

Bessemer converter A steel vessel lined with magnesium and calcium oxides. It has air holes in the base and can be tilted.

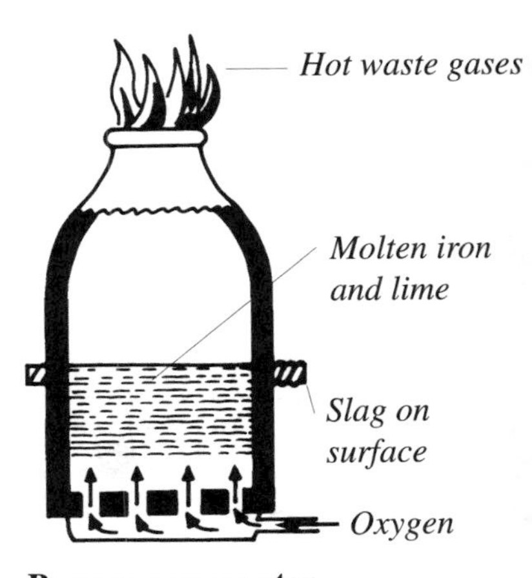

Bessemer converter

Bessemer process A process by which iron is converted to steel. Molten iron is added to a Bessemer converter while it is tilted to allow oxygen and superheated steam to be blown in. The converter is returned to the vertical position. Silicon, manganese, and carbon impurities burn off; carbon monoxide burns at the mouth of the converter. Phosphorus forms its oxide, which then combines with the lining, forming a basic slag of calcium and magnesium phosphates. Molten steel is tapped off from the base of the converter.

beta particle A beta particle is a high-speed electron emitted by the nucleus of certain radioactive elements during β decay. When a neutron in the nucleus decays to a proton, an electron is emitted, thus the atomic number increases by one. A β ray is a stream of high-energy electrons. They will produce ions in matter through which they pass and will penetrate a layer of several millimeters of aluminum.

Bh Symbol for the element bohrium.

Bi Symbol for the element bismuth.

bimolecular reaction *See* molecularity.

binary compound A compound (such as carbon monoxide, CO) that contains two elements.

biochemistry The branch of chemistry that studies living things.

biodegradable A substance that can be broken down by microorganisms into simpler substances.

biodegradable plastics Plastic with starch incorporated into its structure in order that it can be broken down when it comes into contact with soil.

bismuth Element symbol, Bi; Group 5; brittle reddish white metal; Z 83; A(r) 208.98; density (at 20°C), 9.8 g/cm^3; m.p., 271.3°C; German name *wismut*, in Latin *bisemutum*; discovered 1753; used in low melting alloys; some compounds have medical uses.

Bk Symbol for the element berkelium.

blast furnace A large tower (approximately 100 ft [30 m] high and 20 ft [6 m] wide) used to extract iron from its ores. Iron ore, coke, and limestone are added from the top, and pre-heated air is blown in through tubes (tuyeres) at the base. This causes the coke to burn and leads to several chemical processes, resulting in the reduction of the ore, which settles as a liquid at the base of the tower. A molten slag of calcium silicate floats on this and is removed separately.

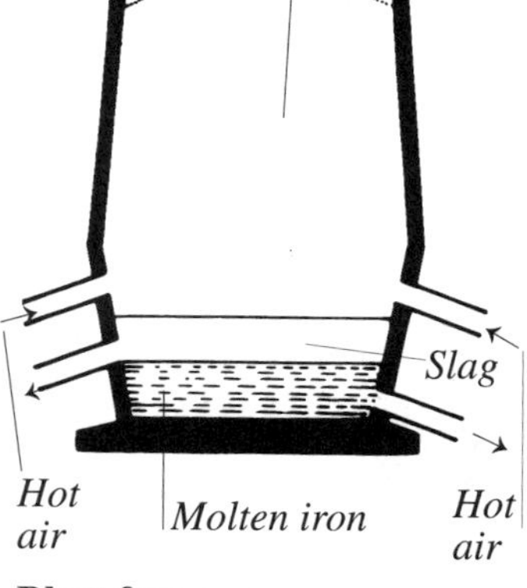

Blast furnace

bleach A substance that can remove the color from another substance, using either an oxidizing agent (such as chlorine) or a reducing agent (such as sulfur dioxide).

bleaching powder A white powder that consists of a mixture of hydrated calcium chloride, calcium hydroxide, and calcium chlorate(I) ($Ca(OCl)_2$) . When treated with a dilute acid, bleaching powder liberates chlorine. Chlorine is the bleaching agent (*see* bleach).

blue-ring test Test for the presence of thiosulfate (a salt containing the ion $(S_2O_3)^{2-}$). Take two test tubes. Pour 3 ml concentrated sulfuric acid in the first. In the second test tube add a small sample of the substance being tested to about 5 ml of aluminum molybdenate solution and shake to mix. While holding the first test tube at an angle of 45°, carefully pour some liquid from the second tube to form two liquid layers. If a deep blue ring is seen forming at the sulfuric acid/solution border, the solution contains a thiosulfate salt.

blue vitriol Hydrated copper sulfate $CuSO_4.5H_2O$. (Also known as copper sulfate pentahydrate.) Copper sulfate in this form exists as blue crystals.

bohrium Element symbol, Bh; transition element; Z 107; A(r) 262; named in honor of Danish physicist Niels Bohr; discovered 1981. Formerly known as unnilseptium.

boiling The process by which a substance changes from the liquid state to the gas state at a fixed temperature (the boiling point). At this point the vapor pressure of the liquid equals that of the atmosphere.

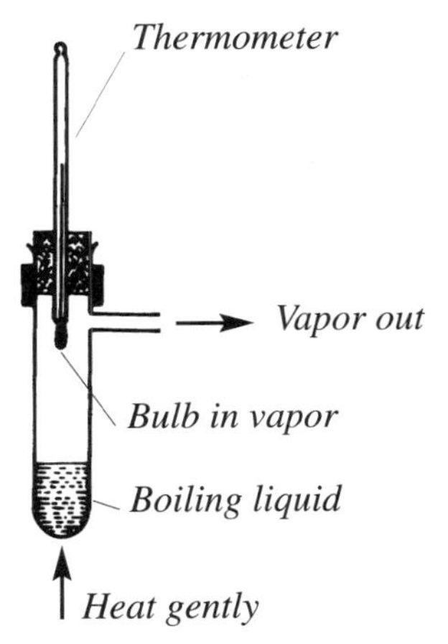

Boiling point

bond A bond is the chemical connection between atoms within a molecule. Bonds are forces and are caused by electrons. Covalent bonds form when two electrons are shared between two atoms (usually between two nonmetallic atoms), one contributed by each atom. Covalent double bonds form when four electrons are shared between the two atoms. Covalent triple bonds form when six electrons are shared between the two atoms. Coordinate bonds are a type of covalent bond and form when one of the atoms supplies both electrons. Ionic bonds (electrovalent or polar bonds) form when atoms form ions and electrons are transferred from one atom to another. The ions are held together by electrostatic attraction. *See* metallic bond.

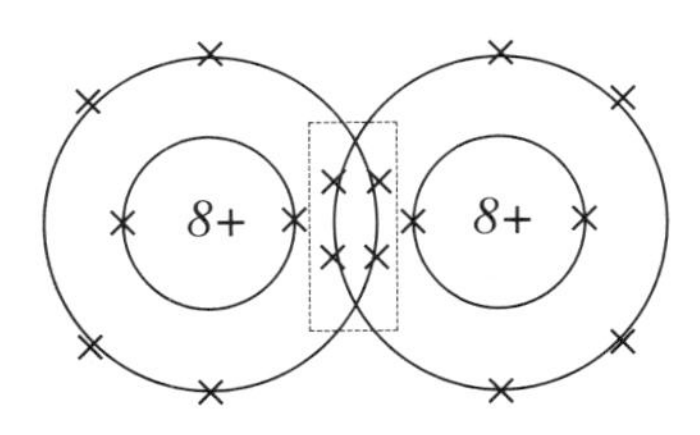

Bond + double covalent bond in oxygen

bond energy During a chemical reaction, bonds between some of the atoms present are broken and new bonds are made. When bonds are broken, energy is absorbed; when bonds are formed, energy is evolved. The energy change in the reaction is the energy of a bond. Bond energies of multiple bonds are usually greater than those of single bonds. The energy of the hydrogen bond may be thought of as the energy absorbed when one mole of hydrogen molecules is split into free atoms. Bond energies can be calculated from the standard enthalpy of formation of the compound and from the enthalpies of atomization of the elements. Bond energies give the energy required to break the bonds and are hence a measure of the relative stabilities of the bonds.

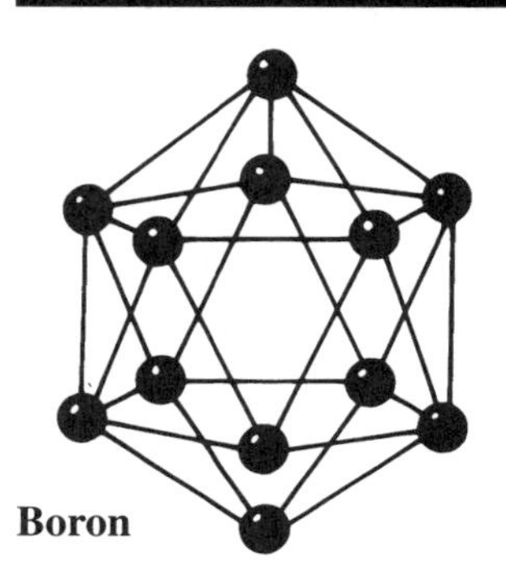

Boron

bonding orbital *See* molecular orbitals.

borax $Na_2B_4O_7.10H_2O$ (disodium tetraborate-10-water, sodium tetraborate). A naturally occurring sodium salt. It is used in the glass industry and as a mild antiseptic.

boron Element symbol, B; group 3; very inert; Z 5; A(r) 10.81; density (at 20°C), 2.35 g/cm^3; m.p., 2079°C; name derived from the Arabic *burak*; discovered 1808; used in nuclear reactors. Boron filaments are used in epoxy resins. Most of the boron used is in borosilicates in enamels and glasses.

boron carbide B_4C. A black solid that is very hard (9.5 on Mohs' scale). It is used as an abrasive.

boron nitride BN. A very hard solid that is insoluble in cold water. It sublimes above 3000°C. It has high electrical resistance and high thermal conductivity and is used in the electrical industry.

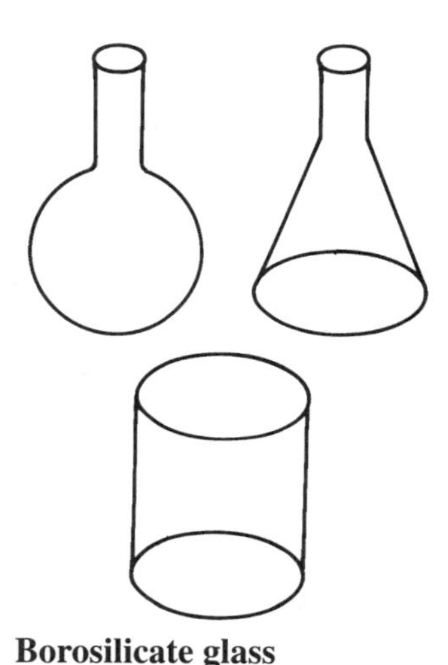

Borosilicate glass

borosilicate glass Glass made by the addition of boron oxide (B_2O_3) to the normal silicate network of glass. It forms a glass (such as Pyrex) that has a low coefficient of thermal expansion that allows it to be exposed to rapid heating or cooling without cracking.

borosilicates Substances in which BO_3 and SiO_4 are linked to form networks that have many structures.

Bosch process The production of hydrogen from water gas by passing a mixture of water gas and steam over an iron catalyst at about 500°C. $CO + H_2O = CO_2 + H_2$. The carbon dioxide is removed by washing with water or with potassium carbonate solution under pressure. $K_2CO_3 + CO_2 + H_2O = 2KHCO_3$.

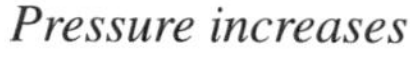

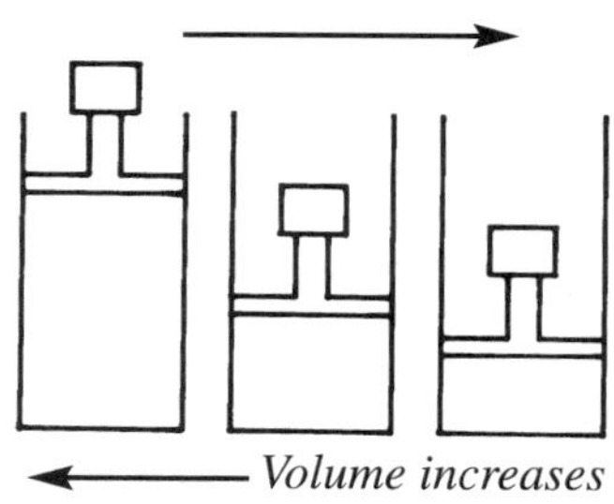

Boyle's law

Boyle's law The volume of a given mass of gas varies inversely with its pressure at constant temperature. One of the three ideal gas laws.

Br Symbol for the element bromine.

branched chains A line of carbon atoms having side groups attached to the chain.

brass An alloy of copper and up to 40% of zinc. It is harder and has more corrosion resistance than copper. It is used for electrical components and ornaments.

breeder reactor A nuclear reactor that produces more material capable of nuclear fission than it consumes.

brine A strong solution of sodium chloride in water.

bromides Compounds derived from hydrobromic acid (HBr). Silver bromide is used in photography and some bromides are used medicinally as sedatives.

bromine Element symbol, Br; halogen, group 7; dark red liquid, vapor is red and poisonous; Z 35; A(r) 79.9; density (at 20°C), 3.12 g/cm^3; m.p., –7.2°C; very reactive oxidizing agent; name derived from the Greek *bromos*, "stench"; discovered 1826; used to make ethylene dibromide and manufacture photographic materials, fumigants, water-purifying materials, and flame-proofing agents.

bromine test To test for an unsaturated hydrocarbon. Add bromine solution (orange in color) to the hydrocarbon being tested. If the hydrocarbon contains unsaturated bonds, the bromine solution is decolorized. This test uses the ability of bromine molecules to add on to a double bond, forming a colorless halocarbon compound.

bronze An alloy of copper and tin (less than 10%). It is much stronger than copper and its discovery was important in the history of human civilization. Its uses now are in gear wheels and engine bearings.

brown-ring test The chemical test for the presence of nitrates. The sample is dissolved in water in a test tube. A solution of iron(II) sulfate is added and the two solutions are mixed. Concentrated sulfuric acid is added slowly so that it sinks to form a layer beneath the aqueous solution. If nitrate is present in the sample, a brown ring is formed at the junction between the sulfuric acid and the aqueous solution. This disappears if the tube is shaken.

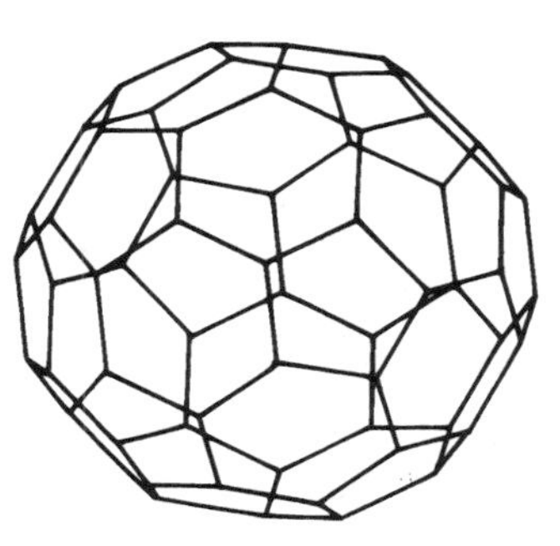

Buckyball molecule

buckyball molecule or **buckminsterfullerene** An allotropic form of carbon. It has a cage-like structure and has the formula C50, C60 and C70.

buffer solution A solution that can maintain an almost constant pH value when dilute acids or alkalis are added to it. It is made up of a dilute acid or base with a solution of one of its salts and can "mop up" excess hydrogen ions from acids or excess hydroxide ions from bases, maintaining a constant pH. Buffers are present in body fluids such as blood.

bunsen A burner used in the laboratory. It burns a variable mixture of gas and air, the proportions of which can be changed by changing the air hole on the side of the burner.

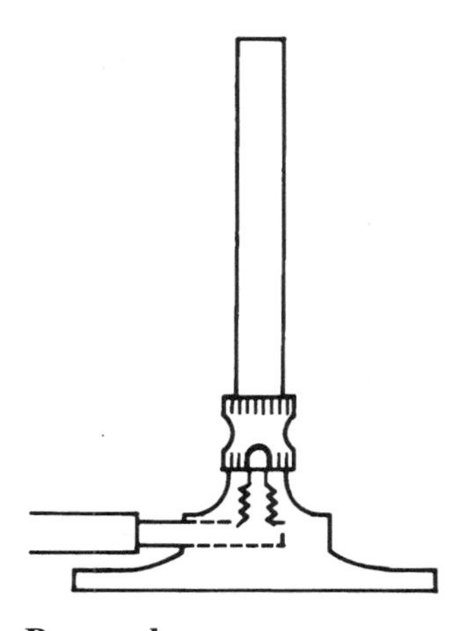

Bunsen burner

burette A long, graduated glass tube with a tap at the lower end. It is used to measure a volume of liquid accurately.

burning *See* combustion.

butane C_4H_{10}. A flammable, colorless gas with a slight smell; m.p., –138.4°C; b.p., –0.5°C. It is a saturated hydrocarbon belonging to the alkane homologous series. Butane is used as a fuel. It is isomeric with 2-methylpropane (formerly called isobutane). $CH_3CH(CH_3)CH_3$.

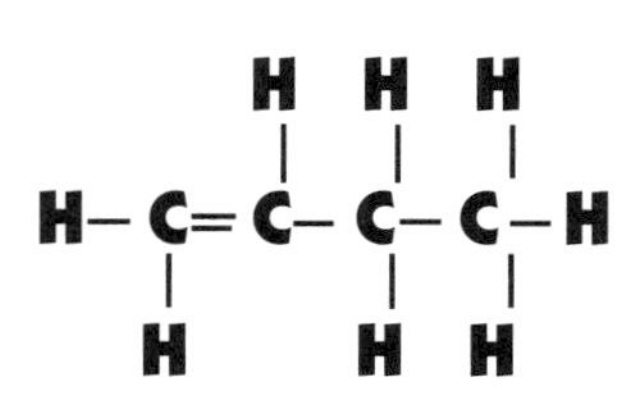

Butene

butanol C_4H_9OH. An aliphatic alcohol with four isomeric forms:

1-butanol, $CH_3CH_2CH_2CH_2OH$, b.p., 117°C.

Isobutanol or 2-methylpropanol, $(CH_3)_2CHCH_2OH$, b.p., 108°C.

Secondary butyl alcohol or 2-butanol, $CH_3CH_2CH(CH_3)OH$, b.p., 100°C.

Tertiary butyl alcohol or 2-methyl-2-propanol, $(CH_3)_3COH$, which occurs as colorless prisms, m.p., 25°C, b.p., 83°C.

butene C_4H_8. An unsaturated hydrocarbon belonging to the alkene homologous series. Three isomers are possible:

1-butene, $CH_3CH_2CH=CH_2$

2-butene, $CH_3CH=CHCH_3$

isobutene or 2-methylpropene$(CH_3)_2C=CH_2$

They are all normally colorless gases (b.p. between –6°C and +3°C) with unpleasant odors.

byproduct A substance produced in a reaction in addition to the required product. (Slag is produced as a byproduct of iron manufacture.)

C Symbol for the element carbon.

Ca Symbol for the element calcium.

cadmium Element symbol, Cd; transition element; white shiny metal; Z 48; A(r) 112.41; density (at 20°C), 8.65 g/cm^3; m.p., 320.9°C; compounds very toxic; name derived from the Greek *kadmeia*, "calamine," from Cadmus (founder of Thebes); discovered 1817; used for electroplating and in alloys. Compounds used in pigments and in color TV tubes.

Rain

Chalk

Rocks

Streams containing dissolved solids

Chalk (calcium carbonate) dissolves in rainwater

calcium Element symbol, Ca; alkaline earth metal, group 2; soft silvery white metal; Z 20; A(r) 40.08; density (at 20°C), 1.54 g/cm^3; m.p., 839°C; name derived from the Latin *calx*, "lime"; discovered 1808; used as a reducing agent and as a getter.

calcium carbide *See* calcium dicarbide.

calcium carbonate $CaCO_3$. A white solid that is sparingly soluble in water. It forms calcium oxide and carbon dioxide when heated. Calcium carbonate occurs naturally in marble, limestone, chalk, and calcite. It dissolves in dilute acids (in rainwater that is used as public water supply, this causes temporary hardness). It is a raw material in the Solvay process and is also used in manufacture of lime (CaO), cement, and glass.

calcium chloride $CaCl_2$. This exists as an ionic compound (Ca_2^+ $2Cl^-$). It is nonvolatile and soluble in water. Its aqueous solution is an electrolyte.

calcium dicarbide CaC_2(calcium carbide, carbide) A colorless solid; r.d., 2.22; m.p., 450°C; b.p., 2300°C. It is produced industrially by a reaction between coke and calcium oxide (CaO) at a temperature of about 2000°C in an electric furnace. Ethyne (C_2H_2) is produced when water is added to calcium dicarbide.

calcium hydrogencarbonate $Ca(HCO_3)_2$ (calcium bicarbonate) This is only stable in aqueous solution. It is formed in nature when water containing carbon dioxide (rainwater) attacks rocks containing calcium carbonate. The insoluble calcium carbonate forms soluble calcium hydrogencarbonate. $CaCO_3 + CO_2 + H_2O = Ca(HCO_3)_2$. Calcium hydrogencarbonate forms temporary hardness in water because, when heated, the insoluble carbonate is formed and is precipitated in vessels used to boil water.

calcium hydroxide $Ca(OH)_2$ (slaked lime) A white powder that dissolves sparingly in water. It absorbs carbon dioxide to form calcium carbonate. It is manufactured by adding water to calcium oxide, the process is known as slaking and is highly exothermic. $CaO + H_2O = Ca(OH)_2$.

calcium nitrate $Ca(NO_3)_2$. A white deliquescent compound; r.d., 2.5; m.p., 561°C. It is very soluble in water. It is formed by reacting nitric acid with a calcium salt (oxide, carbonate or hydroxide). The tetrahydrate form ($Ca(NO_3)_2.4H_2O$) can be crystallized from a solution of calcium nitrate. The anhydrous form can be obtained from the hydrate by heating, but it decomposes easily to form calcium oxide, nitrogen dioxide and water. It is used as a fertilizer and in the manufacture of explosives and matches.

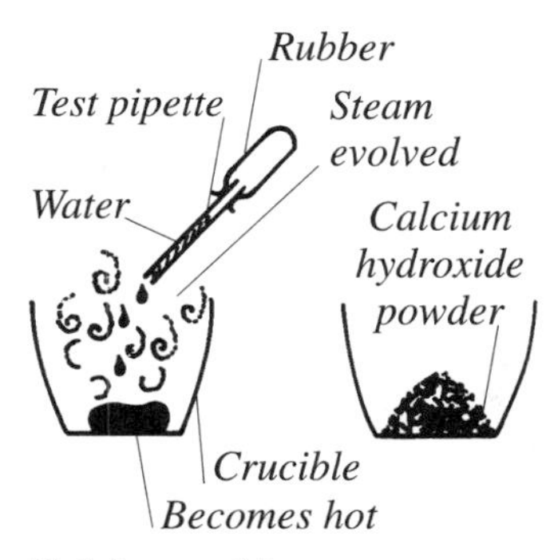

Calcium oxide

calcium oxide CaO (quicklime, lime) A white, hygroscopic powder; r.d., 3.5; m.p., 2600°C; b.p., 2850°C. It has a giant structure and is formed when calcium carbonate is heated strongly. It becomes incandescent at high temperatures. Calcium oxide is used in the manufacture of calcium hydroxide, as a cheap soil conditioner on acid soils, and in the production of iron ore to remove impurities.

calcium phosphate(V) $Ca(PO_4)_2$. A white insoluble powder; r.d., 3.14. It is found in the mineral apatite and is the main component of animal bones. Calcium dihydrogenphosphate(V) and calcium sulfate are formed when calcium phosphate is treated with sulfuric acid. $Ca(PO_4)_2 + 2H_2SO_4 = Ca(H_2PO_4)_2 + 2CaSO_4$. Superphosphate is the name given to this mixture of calcium dihydrogenphosphate and calcium sulfate. It is an important fertilizer.

calcium silicate $CaSiO_3$. It is formed as a slag in a blast furnace during the extraction of iron from iron ore and is used to make cement.

calcium stearate $(C_{17}H_{35}C00)_2$ Ca. An insoluble solid that is formed on the surface of washing water when soap has been used in hard water that contains calcium sulfate ($CaSO_4$).

calcium sulfate $CaSO_4$. A white solid that is sparingly soluble in water (it is a cause of permanent hardness of water). It occurs naturally as anhydrite and (as $CaSO_4.2H_2O$) as gypsum. It is used as a drying agent.

Gypsum, heated at 130°C forms plaster of Paris ($2CaSO_4.H_2O$).

Anhydrite and gypsum are used in the manufacture of sulfuric acid.

californium Element symbol, Cf; actinide; silver-gray metal; Z 98; A(r) 251; named for California; discovered 1950.

calorific value The energy value of a food or fuel, given by the heat produced when a unit amount (1 g or 1 kg) is completely burnt in oxygen.

calorific value of a gas The heat produced by unit volume of the gas when completely burnt.

cane sugar Sucrose ($C_{12}H_{22}O_{11}$) which is extracted in a solution of hot water from crushed sugar cane.

carbide A compound that contains carbon and an element with lower electronegativity. (Compounds containing carbon and oxygen, sulfur, phosphorus, nitrogen or the halogens are not, therefore, carbides, and nor are compounds containing carbon and hydrogen.) Examples of carbides are calcium dicarbide (CaC_2); silicon carbide (SiC); aluminum carbide (Al_4C_3); boron carbide (B_4C). Carbides are formed by heating the components in an electric furnace. Some carbides are very hard. Carbides formed by elements close to carbon in size are covalent, while those formed by highly electropositive elements are ionic.

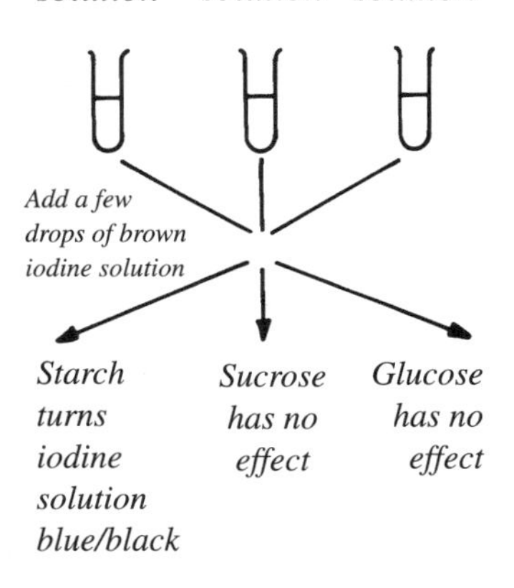

Carbohydrate test

carbohydrate One of a large group of organic compounds that contain carbon, hydrogen and oxygen. They have the general formula $C_x(H_2O)_y$. There are three groups of carbohydrate:

simple sugars (monosaccharides), such as the isomers glucose and fructose ($C_6H_{12}O_6$);

complex sugars (disaccharides), such as the isomers sucrose and maltose ($C_{12}H_{22}O_{11}$);

complex carbohydrates (polysaccharides), such as starch, dextrins and cellulose. Their formulas are $(C_6H_{10}O_5)n$ where n is a large number.

carbohydrates—tests for For a reducing sugar such as glucose (*see* test for reducing sugar). For starch, add iodine solution. If starch is present, the solution will turn blue-black.

carbolic acid *See* phenol.

carbon Element symbol, C; group 4; three isomers, diamond – clear, crystalline; graphite – black, shiny; buckminsterfullerene; Z 6; A(r) 12.01; density (at 20°C) in g/cm^3, 2.25 (graphite), 3.51 (diamond); m.p., 3550°C; name derived from the Latin *carbo*, "charcoal"; known since prehistoric times; active carbon used in industry; carbon 14 isotope(^{14}C) is radioactive and is used in radiocarbon dating. Carbon compounds occur widely in nature, in living organisms, and in fossilized hydrocarbons.

carbonate The carbonate ion CO_3^{2-} has a valency of 2. Group 1 metal carbonates are soluble in water, but all others are insoluble. Carbonates produce carbon dioxide when heated strongly or treated with dilute acid. Thus the test for a carbonate is to add acid and test the resulting gas with limewater. If a carbonate is present there will be a milky precipitate.

carbonation The process of dissolving carbon dioxide in a liquid under pressure. Water is carbonated to make soda water and other fizzy drinks.

carbon bonds Carbon forms four covalent bonds that are arranged symmetrically in three dimensions. *See* tetrahedral compound.

carbon cycle The circulation of carbon through the biosphere. Plants use atmospheric carbon dioxide to make food, which is eaten by animals. Breathing, burning, and decay return carbon dioxide to air.

carbon dating (radiocarbon dating). The way in which the age of previously living animal or vegetable life can be determined. Carbon is present in the atmosphere and in all living tissue in a mixture (the proportions of which are constant while the tissue is living) of isotopes, one of which, ^{14}C, is radioactive with a half-life of 5730 years. When the tissue (animal or vegetable) dies, the proportion of ^{14}C decreases as radioactive decay occurs. The age of a sample of dead material can thus be measured by measuring the radioactivity of the sample.

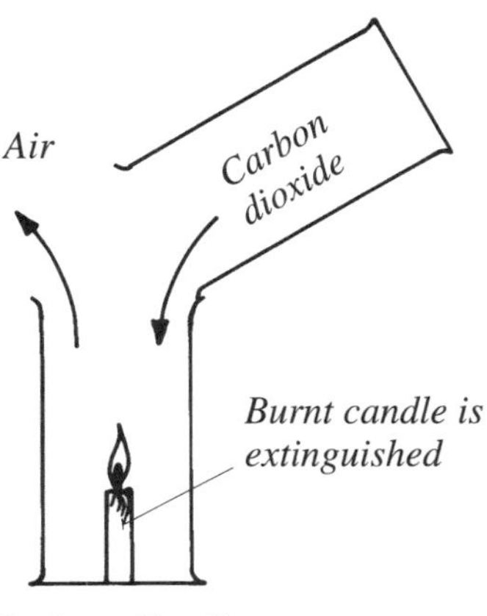

Carbon dioxide

carbon dioxide CO_2. A dense, colorless, odorless gas that does not support combustion; m.p., –56.6, b.p., –78.5. It exists in the atmosphere (0.03%) and is instrumental in the carbon cycle. There is concern that the level of carbon dioxide in the atmosphere is rising and causing global warming (*see* greenhouse effect). Carbon dioxide is soluble in water, forming carbonic acid. Solid carbon dioxide is known as dry ice. To test for the presence of carbon dioxide, pass the gas through limewater; if the gas is carbon dioxide, it forms a white precipitate.

carbonic acid H_2CO_3. A very weak acid formed by dissolving carbon dioxide in water.

carbonization Anaerobic destructive distillation. Coal forms coke in this process; wood forms charcoal.

carbon monoxide CO. A colorless, odorless, very poisonous gas; m.p., –199°C; b.p., –191.5°C. It is sparingly soluble in water and burns in air with a blue flame (this is a test for carbon monoxide). Its toxicity is caused by its ability to bond with hemoglobin in the blood, forming carboxyhemoglobin, which is unable to transport oxygen around the body. Carbon monoxide forms carbonyls with metals because it has vacant *p*-orbitals that are used to form bonds.

carbon tetrachloride *See* tetrachloromethane.

carbonyl group A carbon atom that is attached to an oxygen atom by a double bond and that combines with two other groups of atoms with single bonds.

carboxyl group The organic radical –CO.OH.

carboxylic acid An organic acid that contains one or more carboxyl groups.

carcinogen A substance that can cause cancer.

cast iron Iron obtained from a blast furnace. It contains many impurities, including about 3% of carbon, in addition to phosphorus, silicon, manganese, and sulfur. These impurities make it brittle and it cannot be welded. It is used for objects that are not put under great strain. The Bessemer process and the basic oxygen furnace are two processes of converting cast iron to steel.

Castner-Kellner cell The cell used in the Castner-Kellner process.

Castner-Kellner process The process of electrolysis of brine between graphite anodes and a flowing mercury cathode in a cell.

At the anode the following reactions occur, $2Cl^{-}_{(aq)} = Cl_{2(g)} + 2e^{-}$.

At the cathode, $Na^{+}_{(aq)} + 2e^{-} = 2Na$.

Followed by Na + mercury = amalgam. This amalgam is mixed with water and enters a second cell where the amalgam reacts with water to form hydrogen and sodium hydroxide solution. The mercury is reused. This process was formerly used for the production of sodium hydroxide, used in the chemical industry, but is now more important for the production of chlorine, which is widely used in the manufacture of plastics.

catalysis The alteration of the rate of a chemical reaction because of the presence of a catalyst.

catalyst A substance that alters the rate of a chemical reaction. It takes part in the reaction but remains chemically unchanged by it. Enzymes are the organic catalysts present in animals and plants.

catalytic converter A component of the exhaust system of a car with a gasoline engine. It uses a catalyst of platinum and rhodium to convert various waste products of gasoline combustion (carbon monoxide, nitric oxide, and hydrocarbon compounds that have not undergone complete combustion) to carbon dioxide, nitrogen, and nitrous oxide, thus reducing air pollution.

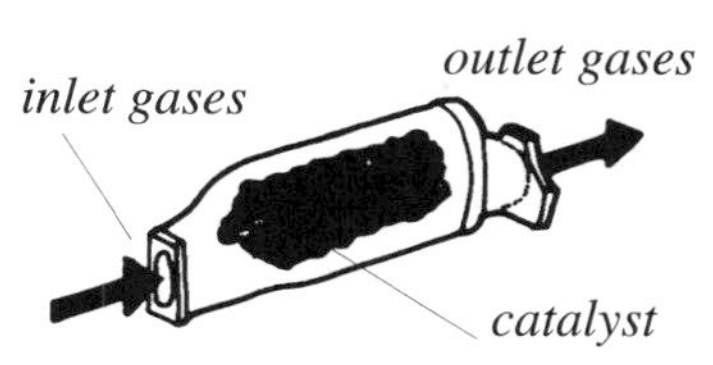

Catalytic converter

catalytic cracking *See* cracking.

catalytic reforming *See* reforming.

cathode The electrode carrying the negative charge in a solution undergoing electrolysis.

cation An ion having positive charge, which is attracted by the negatively charged electrode, the cathode, during electrolysis.

caustic An alkaline substance that burns or corrodes organic material.

caustic potash The common name for potassium hydroxide (KOH).

caustic soda The common name for sodium hydroxide (NaOH).

Cd Symbol for the element cadmium.

Ce Symbol for the element cerium.

cell A vessel, used either to produce electricity or to perform electrolysis, containing an electrolyte in which are dipped two electrodes. There are three main types of cell:

(1) the primary cell, which produces electricity by chemical action (usually irreversible);

(2) the secondary cell, which can be charged by passing electricity through in a direction opposite to the discharge. This reverses the chemical action that produces electricity.

(3) the electrolytic cell in which electrolysis takes place.

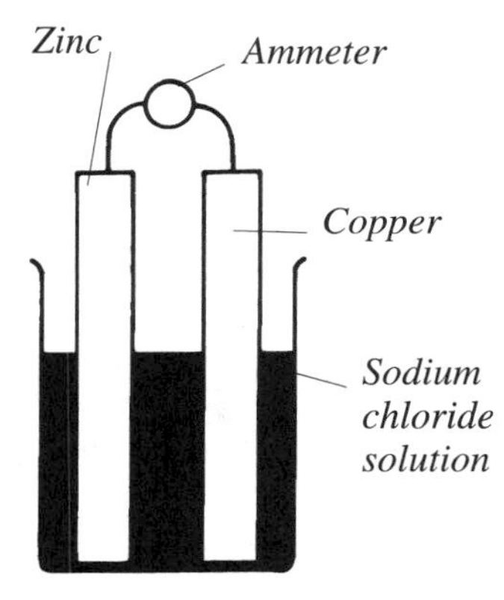

Primary cell

cellulose A complex carbohydrate $(C_6H_{10}O_5)_n$ (*n* is a large number) that is insoluble in water. Cellulose is the main component of the cell walls of plants. Animals can digest cellulose but human beings cannot. Cellulose is used to manufacture paper, cellophane (sheet cellulose manufactured in sheets and used as wrapping material), cellulose ethanoate, and rayon.

cellulose acetate *See* cellulose ethanoate.

cellulose ethanoate (cellulose acetate) A solid flammable substance used in the manufacture of lacquers, magnetic tape, photographic film, and rayon. It is formed by the reaction of cellulose with ethanoic acid using sulfuric acid as catalyst.

Celsius (C) A scale of temperature that has 100 divisions between the lower fixed point (the melting point of pure ice) and the upper fixed point (the boiling point of pure water). 1°=1K.

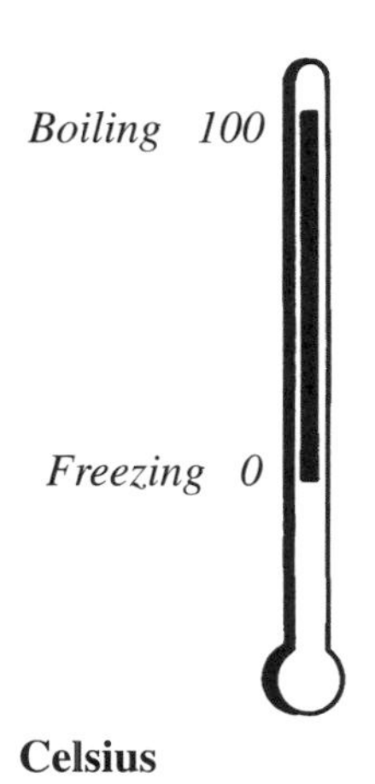

Celsius

cement A gray powder that is a mixture of calcium silicate and calcium aluminate made by heating limestone (calcium carbonate) and clay (containing silicon dioxide and aluminum oxide). A corrosive alkaline mixture is produced when cement is mixed with water. Cement is used as a bonding material in building.

centrifuge A machine that rotates an object at high speed. Under the action of centrifugal force, the rate of sedimentation in a suspension is increased, and particles of different densities can be separated.

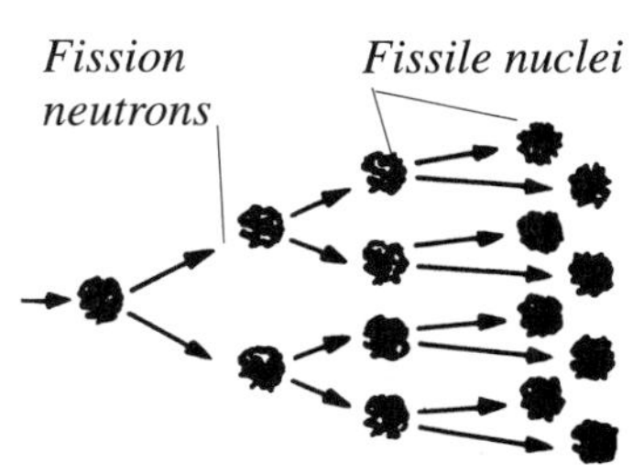

Chain reaction

ceramics Ceramics such as pottery, stoneware, bricks, tiles and pipes are made by shaping clay (a mixture of silica and hydrated aluminum silicate $Al_2O_3.2SiO_2.2H_2O$) into the required form and then firing it in a kiln. This renders it hard, durable and resistant to most chemicals. A surface glaze of a glass (sodium or lead silicate) is usually applied to the object, as it would otherwise be porous. Ceramics are good electrical insulators.

cerium Element symbol, Ce; rare earth/lanthanide; Z 58; A(r) 140.12; density (at 20°C), 6.77 g/cm^3; m.p., 799°C; named for the asteroid Ceres; discovered 1803; used in alloys to improve properties of cast iron and magnesium alloys. Compounds used in ceramic coatings.

cesium Element symbol, Cs; alkali metal, group 1; very reactive metal; Z 55; A(r) 132.91; density (at 20°C) 1.88 g/cm^3; m.p. 28.4°C; name derived from the Latin *caesius*, bluish gray; discovered 1860; used as getter in photoelectric cells (removal of oxygen); isotope 137 used in deep-ray therapy.

Cf Symbol for the element californium.

CFC *See* chlorofluorocarbons.

chain length A measure of the number of atoms linked to form a hydrocarbon chain.

chain reaction A reaction where one event leads to a second, and so on. It is often used to describe a nuclear reaction in which energy is released constantly because neutrons emitted by the fission of an atomic nucleus proceed to cause further fissions, which in turn emit more neutrons.

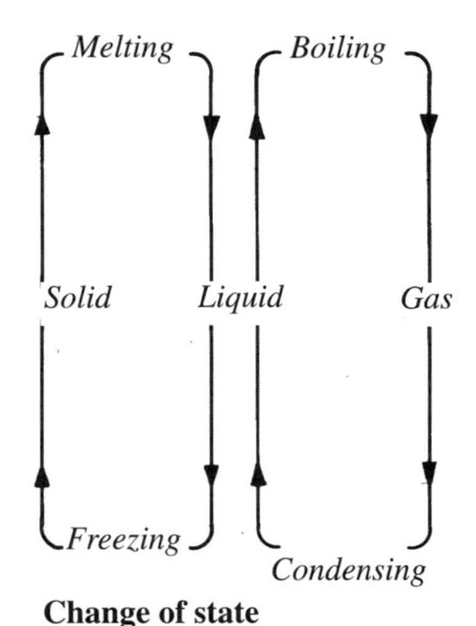

Change of state

chamber process (lead-chamber process) One of the processes used for sulfuric acid production. Sulfur dioxide, oxygen, and nitrogen dioxide react within a large, lead-sheathed brick tower. Sulfuric acid forms as fine droplets that fall to the base of the tower.

change of state The physical process where matter moves from one state to another. Examples of such changes are melting, evaporation, boiling, condensation, freezing, crystallization, and sublimation. A change of state is associated with energy changes.

charcoal The result of the destructive distillation of wood or animal bones. It consists of carbon and has a very open structure with a very large surface area. Gases are easily adsorbed onto the surface of charcoal.

Charles' law The volume (*V*) of a fixed mass of gas at constant pressure (*P*) is dependent on its temperature (*T*).

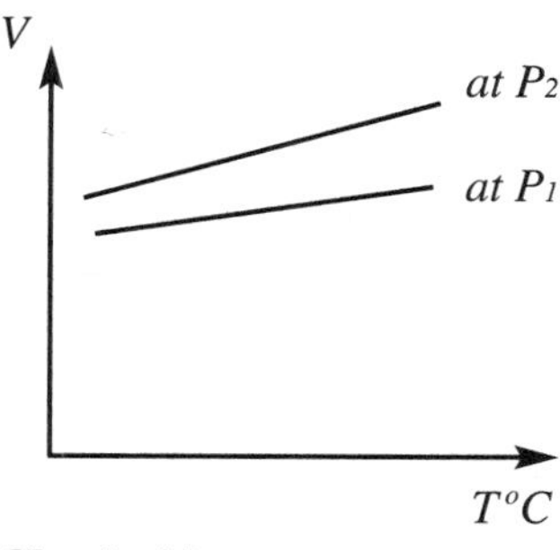

Charles' law

chemical compound A substance composed of two or more elements linked by chemical bonds which may be ionic or covalent. The properties of a compound can be very different from the properties of the elements from which it is made.

chemical energy The energy stored in the bonds between atoms and molecules that is released during a chemical reaction.

chemical equation *See* balanced equations.

chemical equilibrium A chemical reaction that reaches a dynamic equilibrium.

chemical reaction The process in which one or more substances reacts to form new substances. During the process, bonds between atoms are broken and formed as at least one of the original substances is changed to another.

chemiluminescence Light radiated during a chemical reaction.

Chile saltpeter A naturally occurring compound containing sodium nitrate ($NaNO_3$) and some sodium iodate ($NaIO_3$). There are large deposits in Chile. It is used as a fertilizer and in nitric acid manufacture.

china clay A white powder composed of complex aluminum salts used in manufacture of pottery and as a filler in textiles and paper. It is also known as kaolin and is a very pure form of clay.

chlorides Compounds containing chlorine and another element. If the element combined with chlorine is a nonmetal, such as carbon or hydrogen, its chloride is a covalent compound and will be either a liquid with a low boiling point or a gas. If the element is a metal, its chloride will be an ionic solid. Silver nitrate is used to test for the presence of a chloride. If a white precipitate is formed on mixing a solution of a compound with silver nitrate solution and the precipitate dissolves in ammonia solution, the compound being tested contains a chloride.

chlorination Term refers to two processes. **1.** The use of chlorine to disinfect water used for drinking or in swimming pools. **2.** Reactions introducing one or more chlorine atoms into a hydrocarbon structure to form a chlorinated hydrocarbon (*see* halogenation).

chlorine Element symbol, Cl; halogen, group 7; greenish poisonous gas; Z 17; A(r) 35.45; density (at 20°C), 3.214 g/l at STP; m.p., –101°C; powerful oxidizing agent; name derived from the Greek *khloros*, "green"; discovered 1774; used widely in chemical industry in manufacture of chlorinated hydrocarbons; also used in water sterilization and bleaching compounds.

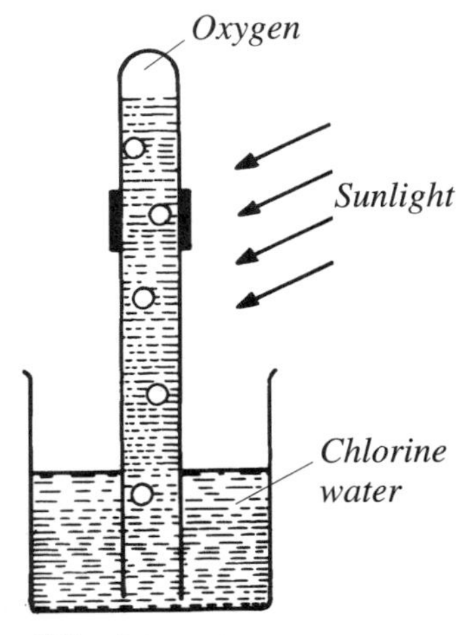

Chlorine water

chlorine—isotopes Chlorine has two isotopes. Chlorine-35 contains 18 neutrons and 17 protons in its nucleus; chlorine-37 contains 20 neutrons and 17 protons in its nucleus. Chlorine gas contains approximately three times more chlorine-35 than chlorine-37; this gives chlorine the relative atomic mass of approximately 35.5.

chlorine water A yellow solution made by passing chlorine gas into ice-cold water. The water absorbs about two and a half times its volume of gas. Chlorine water is a mixture of hydrochloric acid (HCl) and hypochlorous acid (HClO).

chloroethene C_2H_3Cl (also known as vinyl chloride). A gas with m.p. –153.8°C and b.p. –13.37. It is made by chlorinating ethene to form dichloroethane and then removing hydrogen chloride. It is the monomer from which polychloro(ethene) (formerly polyvinyl chloride or PVC) is made.

chlorofluorocarbons Compounds formed when some or all of the hydrogen atoms in a hydrocarbon (typically an alkane) have been replaced with chlorine and fluorine.They are inert substances that have been used widely as refrigerants and as propellants in aerosol cans. Their use is being discontinued as they have been implicated in the destruction of the ozone layer above the Earth and have contributed to the greenhouse effect.

chloroform *See* trichloromethane.

chlorophyll A green pigment normally found in plant leaves. It traps energy from the Sun, which is used by the plant to form glucose by photosynthesis.

chromatography A way of separating and identifying mixtures of solutes in a solution. The method depends on the affinity of the different solutes in the mixture for the medium through which solution moves.

chromium Element symbol, Cr; transition element; hard silvery white metal; Z 24; A(r) 52; density (at 20°C), 7.2 g/cm^3; m.p., 1857°C; very resistant to oxidation; name derived from the Greek *khroma*, "color"; discovered 1797; used extensively as a steel additive and for electroplating.

chromophore A group of atoms responsible for the color of a compound – the azo group is a chromophore.

citric acid $C_6H_8O_7$. A white crystalline solid. It is a weak organic acid that contains three carboxyl groups and one hydroxyl group. Citric acid is found in the juice of lemons and some other fruits.

Cl Symbol for the element chlorine.

clay A fine-grained deposit formed by weathering of rocks. It is mainly composed of hydrated aluminum silicates and usually contains some impurities, such as iron, calcium, and magnesium oxides. Very pure clay is white (*see* china clay).

Cm Symbol for the element curium.

Co Symbol for the element cobalt.

coagulation The grouping together of small particles in a solution into larger particles. Such a solution eventually coagulates with the particles forming either a precipitate or a gel.

coal A fossil fuel containing (approximate percentages) carbon, 80%; oxygen, 8%; hydrogen, 5%; and sulfur, 1%, with some nitrogen and phosphorus.

coal gas A mixture of hydrogen, methane, and carbon monoxide produced by the destructive distillation of coal.

coal tar One of the products of the destructive distillation of coal. It is a black liquid containing hundreds of organic compounds (such as benzene, toluene, naphthalene, and phenol), which can be separated by fractional distillation. Coal-tar derivatives are important in the manufacture of dyes, drugs, insecticides, and other organic chemicals.

cobalt Element symbol, Co; transition element; silvery white metal; Z 27; A(r) 58.93; density (at 20°C), 8.83 g/cm^3; m.p., 1495°C; name derived from the German *kobold*, "goblin"; discovered 1735; used in alloys.

cobalt chloride $CoCl_2$. Its anhydrous form is blue and its hydrated form is pink. Anhydrous cobalt chloride is used to test for the presence of water.

coenzyme A small organic nonprotein molecule that acts with an enzyme in many enzyme-catalyzed reactions.

coke The solid residue produced by the destructive distillation of coal.

colloid A substance made of very small particles whose size (1–100 nm) is between those of a suspension and those in solution.

combining mass *See* equivalent mass.

combining power (valency) *See* valency.

combustion The chemical term for burning, usually in oxygen.

common salt *See* sodium chloride.

complex ion A cation formed when an atom or group of atoms (*see* ligand) donate electrons to form coordinate bonds with a metal ion or atom.

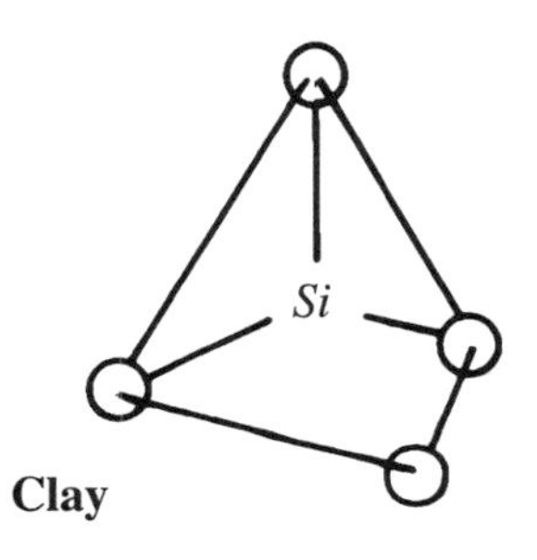

Clay

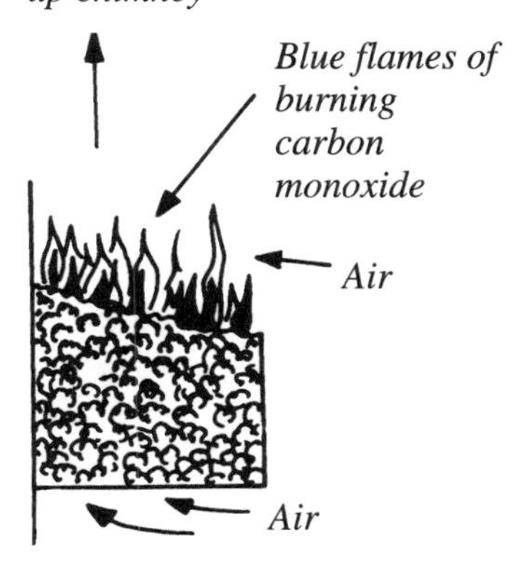

Coke

Many complex ions are formed by transition metals because they are able to accept the donated electrons. The ammonium ion (NH_4^+) and the hydroxonium ion (H_3O^+) are also complex ions.

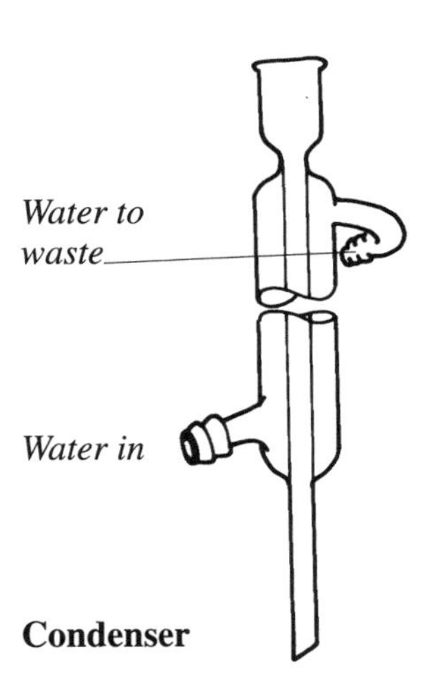

Condenser

compound *See* chemical compound.

concentration A measure of the quantity of solute dissolved in a solution at a given temperature. Units used are grams of solute per liter of solution, molarity, and percentage.

concrete A mixture of cement with sand and gravel. It sets to a rock-like mass when mixed with water because the silicates and aluminates in the cement form long thread-like crystals when hydrated.

condensation The process by which a liquid forms from its vapor.

condensation reaction The joining together of two or more molecules with the elimination of a small molecule (usually water).

condensation polymerization A process by which molecules join together in a series of condensation reactions. When molecules join together in this way, a small molecule (usually water) is eliminated and larger molecules, or macromolecules, are formed that consist of repeated structural units.

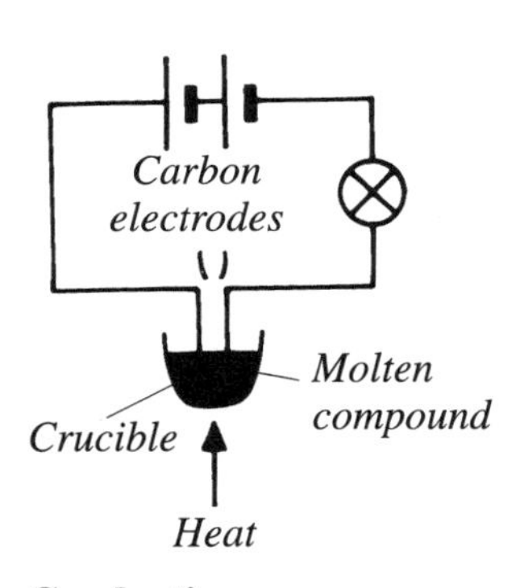

Conduction

condenser An apparatus in which a vapor is converted to a liquid. In a condenser (Liebig condenser), the tube through which the vapor flows is surrounded by a jacket in which water flows.

conduction (1) (*electrical*) The movement of free electrons from atom to atom in a metallic conductor, which transfers electrical energy. The current (flow of charge per second) depends on the circuit's resistance (Ohm's law). (2) (*thermal*) *see* thermal conduction.

conductor A material that is able to conduct heat and electricity.

conformation A particular three-dimensional shape taken by a molecule. Many shapes are possible, given that part of the molecule can rotate about a single bond.

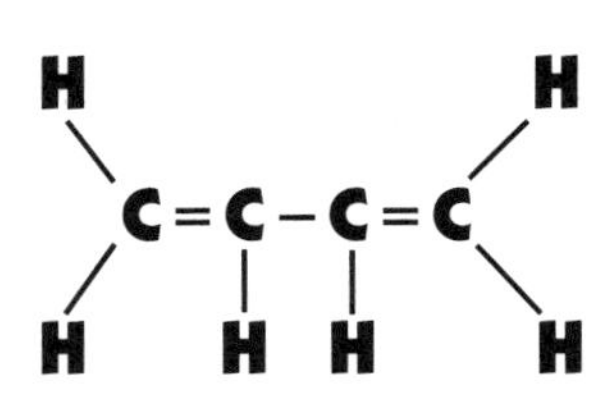

Conjugated structure

conjugated structure A structure that has alternate single and double (or triple) bonds between carbon atoms in an organic compound.

conjugate solutions Solutions of two substances that are partially miscible will form two conjugate solutions in equilibrium at a certain temperature.

constant boiling mixture *See* azeotrope.

contact process The industrial process used to manufacture sulfuric acid. It uses iron pyrites.

control experiment (control) An experiment that is performed at the same time as an experiment investigating the operation of a particular factor. In the control experiment this factor remains constant in order that the effect of the particular factor may be studied.

coordinate bond *See* bond.

copolymer A polymer formed by the polymerization of more than one monomer.

copper Element symbol, Cu; transition element; pinkish metal; Z 29; A(r) 63.55; density (at 20°C), 8.92 g/cm^3; m.p., 1083.4°C; brightly colored salts; name derived from the Latin *cuprum*; known from prehistoric times; used widely in alloys (brass, bronze); used in wire and piping; compounds used in pigments, paints, and fungicides.

copper(II) carbonate Its formula is $CuCO_3$, but it is unknown in this state. It occurs as $CuCO_3.Cu(OH)_2$, a green insoluble solid. It is soluble in both dilute acids and ammonia solution. It decomposes to form copper oxide, carbon dioxide, and water vapor when heated.

copper chlorides CuCl (copper(I) chloride) A white insoluble solid that is formed by boiling copper with copper(II) chloride solution and concentrated hydrochloric acid. The solution is then poured into water.

$CuCl_2$ (copper(II) chloride) An anhydrous soluble brown solid. A concentrated aqueous solution of copper(II) chloride is brown. The color of the solution changes to green ($CuCl_2.2H_2O$), then blue as more water is added.

copper(II) hydroxide $Cu(OH)_2$. A blue-green insoluble gelatinous base that decomposes to form copper(II) oxide and water vapor when heated. It is formed by the action of an aqueous solution of a copper(II) salt with sodium hydroxide.

copper(II) nitrate $Cu(NO_3)_2$. (Usually, $Cu(NO_3)_2.3H_2O$). A blue, deliquescent soluble salt that decomposes to form copper(II) oxide, nitrogen dioxide, and oxygen when heated.

copper oxides (1) Cu_2O (copper(I) oxide). An insoluble red solid that is made by reducing copper(II) sulfate solution. (2) CuO_2 (copper(II) oxide) An insoluble black solid obtained by heating $Cu(NO_3)_2$.

copper plating To plate an item with copper, it should be thoroughly cleaned, then immersed in a solution of copper sulfate solution. A copper rod is also placed in the solution, and the item to be plated is connected to an electrical source together with the copper rod (the copper rod being the anode and the item to be plated the cathode). If the item is rotated in the solution while a small current flows, it will be coated evenly with copper.

copper pyrites The copper ore $CuFeS_2$. To extract copper, the ore is roasted in air to form a molten mixture of copper(I) sulfide and iron(II) oxide.

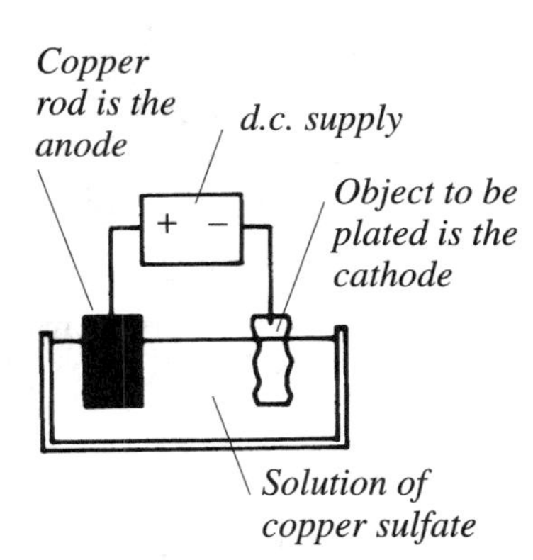

Copper plating

This is heated with sand, and the iron(II) oxide forms a silicate slag. Some of the copper(I) sulfide forms copper(I) oxide and this reacts with the remaining copper(I) sulfide to form copper.

copper(II) sulfate $CuSO_4.5H_2O$. A blue, soluble salt that can be formed by the action of hot concentrated sulfuric acid on copper $Cu + 2H_2SO_4 = CuSO_4 + SO_2 + 2H_2O$ or by the reaction between copper(II) oxide and dilute sulfuric acid. $CuO + H_2SO_4 = CuSO_4 + H_2O$. Copper(II) sulfate is used as a wood preservative and as a fungicide and insecticide for plant diseases (in Bordeaux mixture). Anhydrous copper sulfate is white and can be used to test for the presence of water, when it turns blue.

core charge In a molecule having covalent bonds, such as water, where the oxygen nucleus is more massive than the hydrogen nucleus, electrons in the shared pairs are closer to the oxygen nucleus because of its larger attractive charge than the electrons in the lone pairs.

corrosion The process by which the surface of a metal turns from being an element to being a compound and is thus gradually destroyed. For example, iron corrodes to form rust (hydrated iron oxide) and the surface of copper becomes green when exposed to the atmosphere. *See* electrical protection, sacrificial protection.

covalency The number of covalent bonds an atom is able to make when forming a molecule.

covalent bond *See* bond.

covalent compounds Compounds consisting of molecules where the atoms in the molecules are held together by covalent bonds. They are liquids and gases with low melting and boiling points.

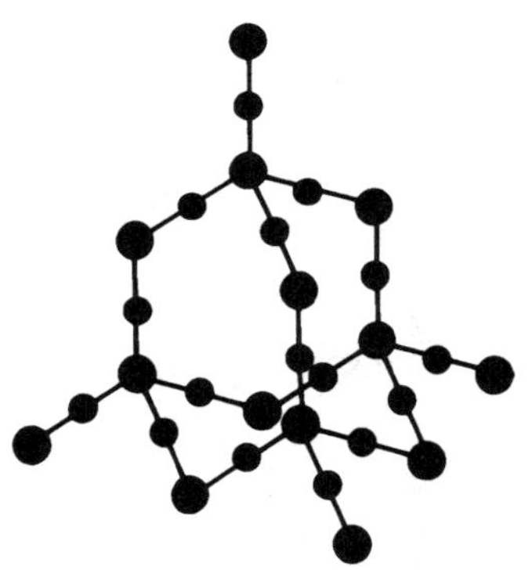

Covalent network structure

covalent network (covalent crystal) A structure in which millions of atoms are linked by single covalent bonds. Such structures have high melting and boiling points.

Cr Symbol for the element chromium.

cracking The process used in the petroleum industry to convert large-chain hydrocarbon molecules to smaller ones. The process uses heat and catalysts.

cream of tartar $C_4H_5O_6K$ (Potassium hydrogen tartarate) A white crystalline solid used in baking powder and medicine.

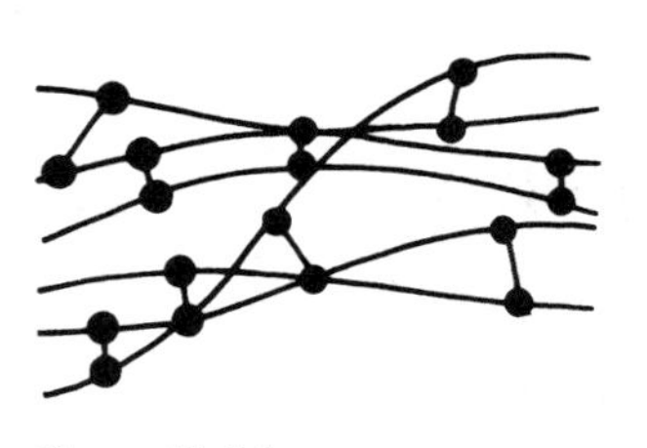

Cross-linking

cross-linking Chemical bonds between adjacent polymer molecules.

crude oil or **petroleum** A mixture of solid, liquid, and gaseous hydrocarbons. It tends to be a thick black liquid that has to be converted to useful products by refining. The different components are separated by fractional distillation, and larger molecules are split into more useful smaller ones by cracking.

cryoscopic constant A constant used in the calculation of freezing-point depression.

crystal A substance with an orderly arrangement of atoms, ions, or molecules in a regular geometrical shape. *See* crystal structure.

crystallization The process of forming crystals from a solution which is concentrated above its saturation point (supersaturated) at a certain temperature.

crystallization—water of *See* water of crystallization.

crystal structure The orderly geometric arrangement, or lattice, of atoms, molecules, or ions in a structure that has a particular regular three-dimensional structure. There are several basic shapes taken by a crystal lattice, depending on the component particles. Shapes can be cubic, tetragonal, rhombic, hexagonal, trigonal, monoclinic, or triclinic. In addition, they can have close-packed structures, in which the shape is said to be face-centered, or more loosely packed, in which case the shape is body-centered.

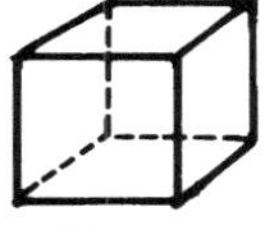

cubic

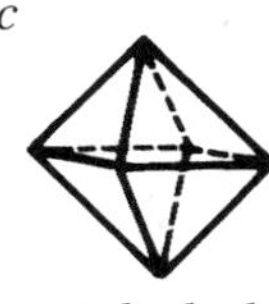

octahedral

tetrahedral

Crystal structure

Cs Symbol for the element cesium.

Cu Symbol for the element copper.

curium Element symbol, Cm; actinide; silvery metal; Z 96; A(r) 247; density (at 20°C), 13.5 (est.) g/cm^3; m.p., 1340°C; rapidly oxidized; named in honor of the scientists Marie and Pierre Curie; discovered 1944.

cyanides Compounds derived from hydrocyanic acid containing the –CN group or the CN^- ion. They are very poisonous.

cycloalkanes Homologous series with the formula C_nH_{2n}. Cycloalkanes have a ring structure and are saturated (they contain no double bonds).

Dacron A polyester fiber made by condensation polymerization between ethane-1,2-diol (ethylene glycol) and the aromatic acid benzene-1,4-dicarboxylic acid (terephthalic acid) $C_6H_4(COOH)_2$. It is used widely in the manufacture of textile fibers, and its texture is similar to that of wool.

Dalton's atomic theory John Dalton, an English schoolmaster, was the first person to formulate a theory of matter. In 1808 he made the following assertions. Matter consists of atoms, which are tiny indivisible particles. Atoms cannot be created or destroyed. The atoms of one element are all identical, particularly in mass, and are different from atoms of other elements. "Compound atoms" (now called molecules) are formed when small numbers of atoms combine chemically. "Compound atoms" within a compound are identical and differ from those of other compounds. Modern atomic theory has superseded this theory.

Dalton's law of partial pressure For a mixture of gases at constant temperature, the total pressure is equal to the sum of the individual pressures (partial pressures) each gas would exert if it were the only gas present in the volume occupied by the entire mixture.

Daniell cell A primary cell where a zinc rod (the negative electrode) is immersed in a saturated solution of either zinc sulfate or dilute sulfuric acid contained within a porous pot. This pot is immersed in a solution of copper(II) sulfate, contained in a copper vessel, which forms the positive electrode. The cell produces an e.m.f. of 1.1 volts.

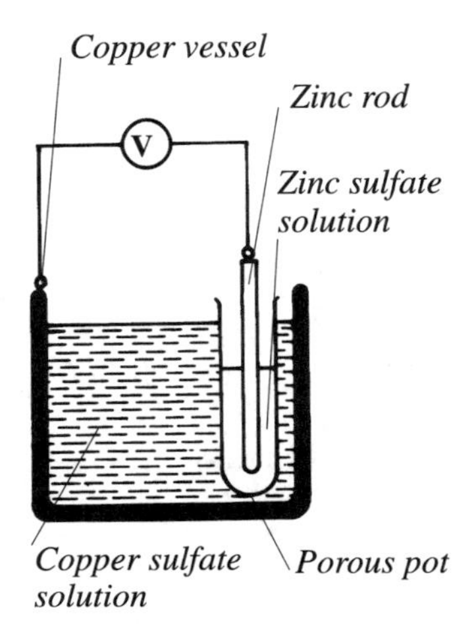

Daniell cell

Db Symbol for the element dubnium.

DDT $(C_6H_4Cl)_2CHCCl_3$ (dichlorodiphenyltrichloroethane) It is a double cyclic organic compound that is insoluble in water. It is a very powerful insecticide and has been successful in controlling malaria. Its use is now restricted as it is not biodegradable and it concentrates in the fatty tissue of animals, where it acts as a poison.

decomposition The process of breakdown of a chemical compound into less complex substances.

dehydrating agent A substance that has an attraction for water and is therefore used as a drying agent. Dehydrating agents can be of different types – a liquid such as concentrated sulfuric acid, a compound such as calcium oxide, which reacts with water to form calcium hydroxide, or an anhydrous salt, which absorbs water.

dehydration A chemical reaction to remove a water molecule from a compound.

dehydrogenation The chemical process of removal of hydrogen atoms from a molecule (a form of oxidation), increasing its degree of unsaturation. For example, the dehydrogenation of ethanol (C_2H_5OH) produces ethanal (CH_3CHO).

deliquescence The way in which a solid substance absorbs water from the atmosphere. The process can continue until the substance passes into solution.

delocalized electron Each atom in a metal has one or more outer electrons that are free to move between atoms. These are delocalized electrons.

depression of freezing point The reduction of the freezing point of a pure liquid when a substance is dissolved in it. The amount of reduction is proportional to the quantity of substance dissolved and not on the type of molecule. The reduction of freezing point ($\Delta t = K_f C_M$ [C_M is the molar concentration of dissolved solute] K_f is the cryoscopic constant).

desiccator A container used to dry and to keep substances dry. It is made of glass or plastic with a close-fitting lid. The substance to be dried is placed on a perforated plate above a dehydrating agent (often silica gel in the laboratory). It can have a tap to remove air.

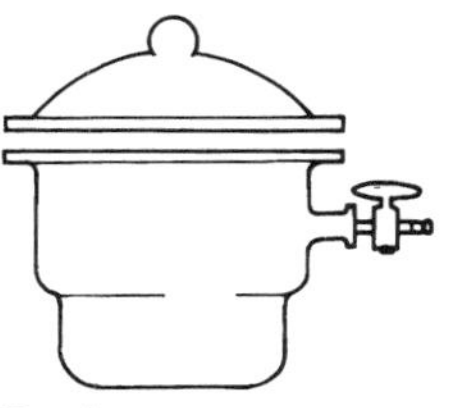
Desiccator

desorption *See* adsorption.

destructive distillation The process of breaking down complex organic substances into a mixture of volatile products in the absence of air. These are condensed and collected.

destructive distillation of coal The products are coke, ammonia liquor, coal tar, and coal gas.

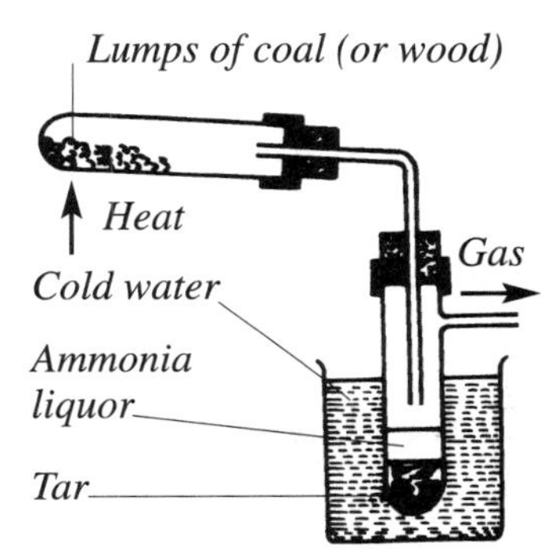

Destructive distillation

desulfurization Removal of sulfur, using hydrogen and a catalyst, from fossil fuels to prevent the release of sulfur dioxide when the fuel is burnt.

detergent The term is usually used for a synthetic soap substitute. Detergents lower the surface tension of water (*see* surfactant), allowing the thorough wetting of objects. They emulsify oils and fats. An emulsion is formed because one end of the detergent molecule is hydrophilic and is attracted to water, the other end, the hydrocarbon chain, is hydrophobic and is attracted to the oil or fat molecules. The oil or fat is thus broken up into small particles and forms an emulsion. The calcium and magnesium salts of detergents are soluble in water and hence do not form a scum. Detergents are made from petroleum.

deuterium An isotope of hydrogen. Its nucleus contains one proton and one neutron and thus has a relative atomic mass of two.

di- A prefix meaning "two."

diamond Naturally occurring transparent, colorless crystalline allotrope of carbon. It is very hard (10 on Mohs' scale). Artificial diamonds are manufactured from graphite (another allotrope of carbon) using intense heat and pressure.

Diamond

diaphragm cell An alternative to the Castner-Kellner method of producing chlorine and sodium hydroxide from brine. In the cell, the anode (a ring of graphite rods) is separated from the cathode (an iron gauze cylinder) by a diaphragm of porous asbestos. Brine passes from the anode, through the diaphragm to the cathode.

Anode reaction $2Cl^{-}2e^{-} \rightarrow Cl_2$.

Cathode reaction $2H_2O + 2e^{-} \rightarrow 2OH^{-} + H_2$. Chlorine gas is released at the anode and hydrogen gas at the cathode where hydroxide ions are also formed.

diatomic molecule A molecule that consists of two like (H_2) or unlike (HCl) atoms.

dibasic acid An acid that has two replaceable hydrogen atoms. A dibasic acid can form both a normal salt (if both hydrogen atoms are replaced) and an acid salt (if only one hydrogen atom is replaced). *See* basicity of acids.

1,2-dibromoethane $BrCH_2.CH_2Br$ (ethylene dibromide) A colorless liquid with a sweet odor; m.p., 10°C; b.p., 132°C. It is used widely in gasoline as an antiknock additive (it combines with the lead from lead tetraethyl).

diesel fuel (diesel oil, gas oil, DERV [diesel engine road vehicle]) A petroleum fuel consisting of alkanes with a chain length of 14–20 in the boiling range 200–350°C. It is used in diesel engines where it is mixed with air and compressed. This mixture explodes.

diffusion The process of rapid random movement of the particles of a liquid or gas that eventually form a uniform mixture.

dilute solution A solution containing a low concentration of solute. To increase dilution, more solvent is added to the solution.

dimer A compound formed by the combination of two identical molecules (monomers). The resulting compound can contain exactly twice the atoms of the monomer. This is addition dimerization. If another molecule (such as water) is formed when two monomers combine to form a dimer, this is condensation dimerization.

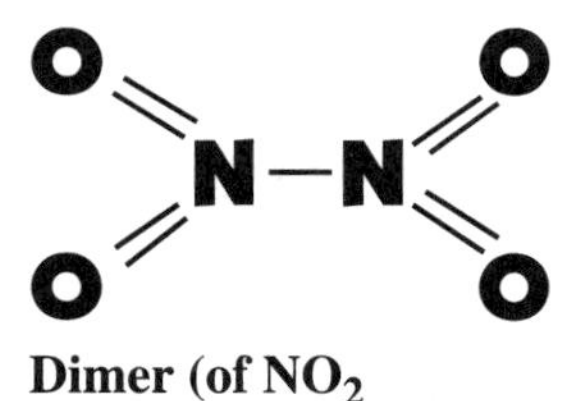

Dimer (of NO_2

dimorphism A substance having two crystal forms.

dinitrogen oxide N_2O (nitrous oxide) (dinitrogen monoxide) A colorless gas with a pleasant smell; m.p., –90.8°C; b.p., –88.5°C. It is moderately soluble in water, forming a neutral solution. It decomposes above 600°C, forming a mixture of nitrogen and oxygen (one-third oxygen by volume). Dinitrogen oxide is used as a mild anesthetic (laughing gas) in dental and other minor operations and as an aerosol propellant.

dinitrogen tetroxide N_2O_4. A colorless solid that melts at 9°C, forming a yellow liquid whose boiling point is 22°C, at which point a brown vapor of nitrogen dioxide (NO_2) is formed.

dipeptide Two amino acids linked by a peptide bond.

disaccharide A sugar molecule formed by a condensation reaction between two monosaccharide molecules (a water molecule is eliminated). Sucrose (a disaccharide) is formed from a molecule of glucose (a monosaccharide) and a molecule of fructose (a monosaccharide) by a condensation reaction. *See* carbohydrate.

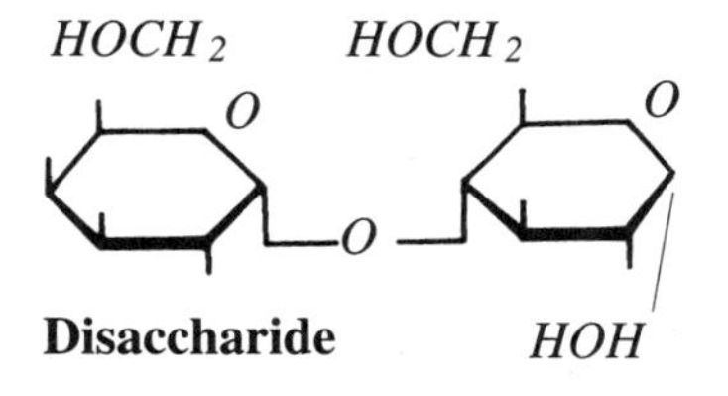

Disaccharide

displacement reaction A metal rod is placed in a solution of a metal salt. The metal rod consists of an element lower in the electrochemical series than the metal ions in solution. The rod will become coated with the ions in solution, and some of the atoms from the metal rod will go into solution as ions. This is a displacement reaction: the ions in the original solution displace the ions in the rod.

dissociation The breaking down of a molecule into smaller molecules, atoms or ions. *See* strengths of acids and bases.

dissolve To add a solute to a solvent to form a uniform solution.

distillation A process in which a solution (or a mixture of liquids whose boiling points are widely differing) is heated to a particular temperature to produce a vapor. This vapor is condensed, forming a pure liquid that has a single boiling point.

divalent Having a valency of two.

d orbital A type of orbital; five types are possible. Each type of d orbital can hold two electrons.

double bond *See* bond, multiple bonds.

double decomposition (metathesis) A chemical reaction between two solutions of ionic salts where an insoluble solid is precipitated. For example, if silver nitrate solution is added to sodium chloride solution, a precipitate of silver chloride is formed. $AgNO_3(aq) + NaCl(aq) \rightarrow AgCl(s) + NaNO_3(aq)$. Some insoluble salts are produced in this way.

Downs cell Chlorine is extracted by electrolysis of molten sodium chloride in a Downs cell. There is a central graphite anode and a cylindrical steel cathode. Chlorine is collected in a hood over the anode.

Dow process Extraction of magnesium from sea water. Magnesium hydroxide is precipitated by adding slaked lime.

dry cell A form of Leclanché cell where the ammonium chloride solution in the zinc casing that forms the negative electrode is replaced by ammonium chloride jelly and the cell is sealed to prevent the electrolyte drying out. (The individual chemicals used may vary between manufacturers.)

dry ice Solid carbon dioxide. Carbon dioxide solidifies at –78.5°C. Dry ice turns directly into a gas (sublimes) if it is heated above this temperature.

drying agent *See* dehydrating agent.

dubnium Element symbol, Db; transition element; Z 105; A(r) 262; named for the Dubna Institute, Moscow; discovered 1967. (Formerly known as hahnium, then unnilpentium.)

Dulong and Petit's law The product of a solid element's relative atomic mass and its specific heat capacity is approximately 6.4. This can be used to find a solid's valency if its equivalent mass has been determined.

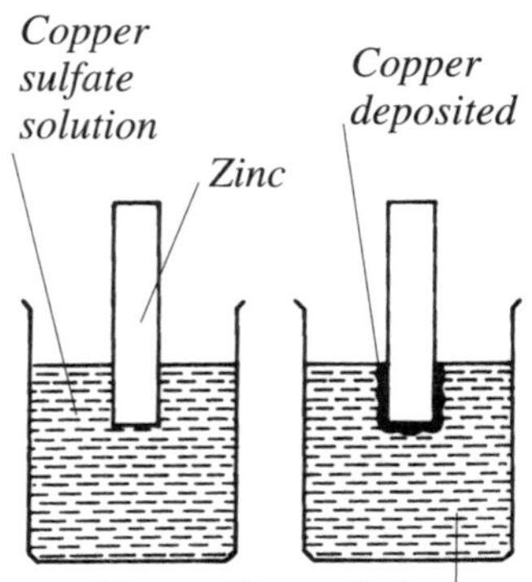

Displacement reaction

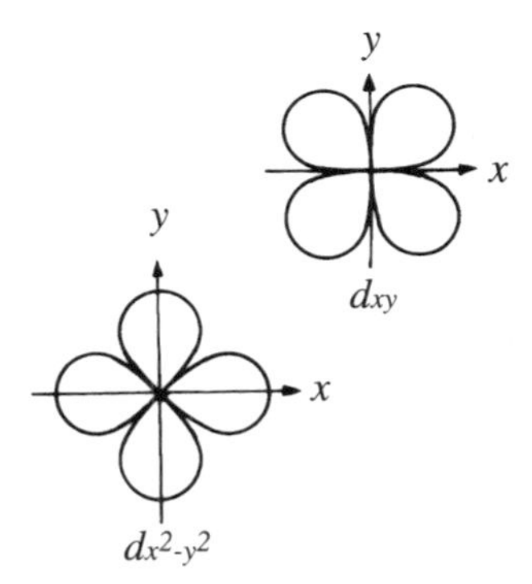

d orbitals

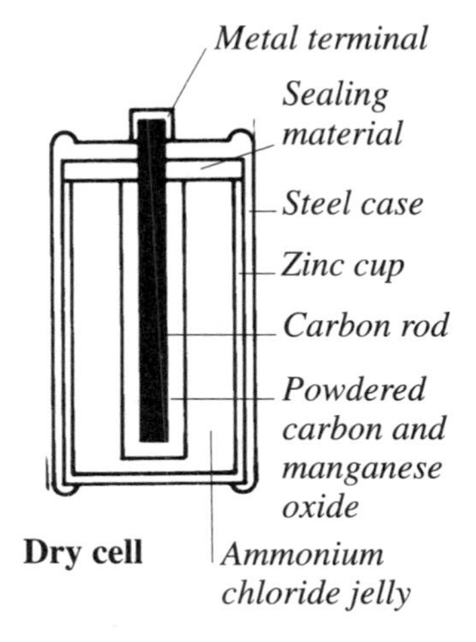

Dry cell

duralumin An alloy of aluminum containing 95% aluminum with copper and magnesium. It is much stronger than pure aluminum.

Dy Symbol for the element dysprosium.

dyes Chemicals, either natural or synthetic (and usually of organic origin), used to color fabrics, paper, plastics, etc. Dyed objects have color because the dye absorbs some of the light falling on it and reflects the rest.

dynamic equilibrium A balanced state of continual change in a system. A reversible chemical reaction may reach a state of dynamic equilibrium when the rate of the forward reaction is equal to the rate of the backward reaction.

dynamite Nitroglycerine (unstable in handling) mixed with a type of clay to produce a stick of dynamite, which is safe to handle and explodes only if detonated.

dysprosium Element symbol, Dy; rare earth element/lanthanide; Z 66; A(r) 162.5; density (at 20°C), 8.55 g/cm^3; m.p., 1412°C; name derived from the Greek *dysprositos*, "hard to get at"; discovered 1886; compounds used in lasers.

effervescence The production of bubbles of gas or air that rise to the surface in a liquid.

efflorescence The way in which a hydrated crystal loses water of crystallization to the atmosphere, making its surface become powdery.

effusion The process by which a gas under pressure moves through a small aperture into a region of lower pressure (*see* Graham's law).

einsteinium Element symbol, Es; actinide; Z 99; A(r) 254; named in honor of Albert Einstein; discovered 1952.

elastomer A substance that can stretch and return to its former shape.

electrical protection Protection of a metal surface against corrosion. A metal loses electrons when it corrodes. If the metal surface is connected to the negative terminal of a direct source of electricity, electrons are supplied to it and corrosion of the surface is inhibited. The bodies of cars and trucks are connected to the negative terminal of the battery to give them some electrical protection.

electrochemical series (electromagnetic series, displacement series, electromotive series) Metallic elements arranged in order of increasing electrode potential (or readiness to release electrons and form cations). The highest elements in the table are the most reactive or electropositive (the most likely to release electrons and form cations).

The series can be used to predict reactions between metals by comparing their positions in the series (*see* displacement reaction).

The values assigned to the elements in the series (large negative values at the top of the table) can be used to calculate the voltage of a cell consisting of two different metal electrodes with an electrolyte between them by using the values of the two metals. The highest voltages are attained when there is a large gap between the two elements in the table.

electrode A conductor that allows current to flow through an electrolyte, gas, vacuum, dielectric, or semiconductor.

electrolysis The process by which an electrolyte is decomposed when a direct current is passed through it between electrodes. Positive cations move to the cathode to gain electrons, negative anions move to the anode to lose electrons. Substances are either deposited or liberated at the electrodes depending on the nature of the electrodes and electrolyte.

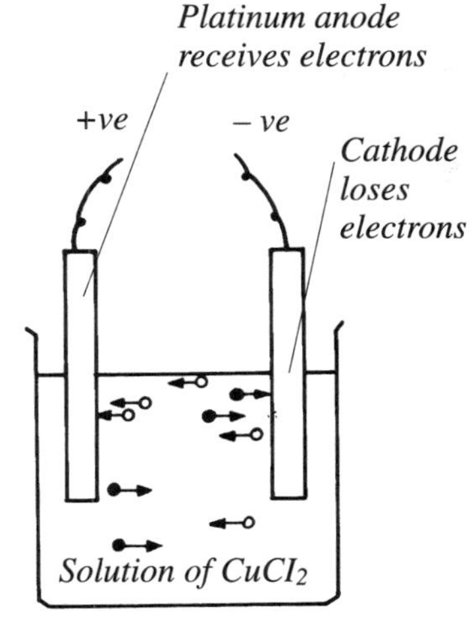

Electrolysis

electrolysis of brine *See* Castner-Kellner cell.

electrolyte A substance that forms ions when molten or dissolved in a solvent and that carries an electric current during electrolysis. Strong electrolytes contains many ions.

electromotive series *See* electrochemical series.

electron One of the three basic subatomic particles. It is very light (its mass is 9.109×10^{-31} kg) and orbits round the nucleus of an atom. It has a negative charge, and in neutral atoms the number of electrons is equal to the number of protons in the nucleus.

electron arrangements *See* electronic structure of atom.

electronegativity A measure of the ease with which an atom can attract electrons. Group 7 of the periodic table contains electronegative elements, fluorine being the most electronegative.

electronic structure of atom (electronic configuration). This gives an indication of the position of electrons around the nucleus of an atom and is useful in showing how the element forms bonds. Electrons are arranged in shells around a nucleus. Each shell can contain a maximum number of electrons, depending on the number of orbitals in the shell (each orbital can contain two electrons). The further from the nucleus a shell is, the more orbitals it can contain. The first shell can contain up to two electrons (in one s orbital), the second up to eight (in one s and three p orbitals), the third up to 18 (in one s, three p and five d orbitals), and the fourth up to 32 (in one s, three p, five d and seven f orbitals). The *nth* shell contains 2^{-2} orbitals.

In the periodic table, elements are arranged in order of increasing atomic number (number of electrons). The shells of electrons around the nucleus of each atom are filled in turn; the Aufbau principle governs the order in which the orbitals are filled. The shell closest to the nucleus is filled first, as it has the lowest energy level. The degree to which these shells are full affects the properties of the element. Elements that have a full outer shell tend to be more stable (the noble gases have a full outer shell). The electronic configuration of each element can be represented. For example, in group 1 of the periodic table (the alkali metal group), lithium is represented as 2.1 (the first shell is full and there is one electron in the second shell); sodium can be represented as 2.8.1 (the first and second shells are full and there is one electron in the third shell). Group 1 elements have an electronic configuration ending in 1, group 2 elements 2, etc.

electrophoresis (cataphoresis) The movement of charged particles, colloidal particles or ions through a fluid under the influence of an electric field. It is a method of analyzing protein mixtures and it can be done using specially prepared paper or on a glass slide coated in a gel.

electroplating Electrolytic coating of a metal with a less reactive one. The metal to be plated is used as the cathode in an electrolyte containing ions of the metal that is used for the plating. These ions are deposited firmly on the surface of the cathode.

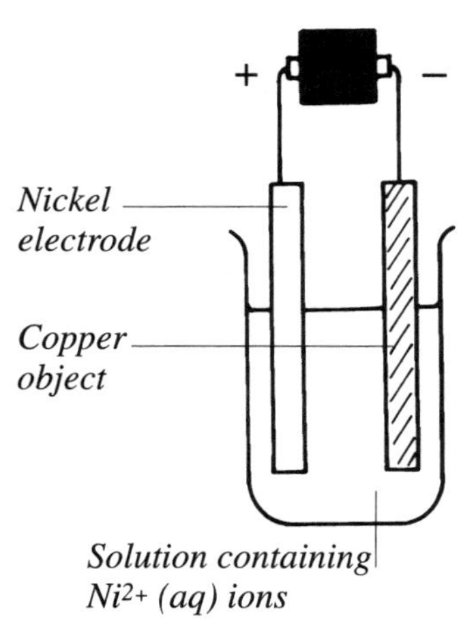

Electroplating

electropositivity A measure of the ease with which an atom loses electrons. Elements from group 1 of the periodic table are all very electropositive.

electrovalent bond *See* bond.

element A substance that cannot be split into simpler substances using chemical methods. An element contains atoms that have the same numbers of protons and electrons (the numbers of neutrons may vary between atoms of an element). One hundred and twelve elements have been discovered, although only 92 occur naturally.

elementary particles The particles from which atoms are made. Neutrons and protons are found in the nucleus of the atom. Electrons form a cloud around the nucleus. *See* fundamental particles.

element—families of Elements in the same group of the periodic table have similar properties – they have the same number of electrons in their outer shell.

elimination reaction A chemical reaction in which an organic molecule loses certain atoms and a double or triple bond is formed. For example, a water molecule can be removed from a molecule of ethanol (using sulfuric acid) leaving a molecule of ethene.

empirical formula The simplest ratio of atoms in a compound.

emulsifier A substance that both assists the formation of an emulsion and stabilizes it when formed.

emulsion A colloidal dispersion of small droplets of one liquid dispersed within another, such as oil in water or water in oil.

enantiotropy The transformation of one allotrope, or form, of a substance into another by a change in temperature. Such a change is reversible.

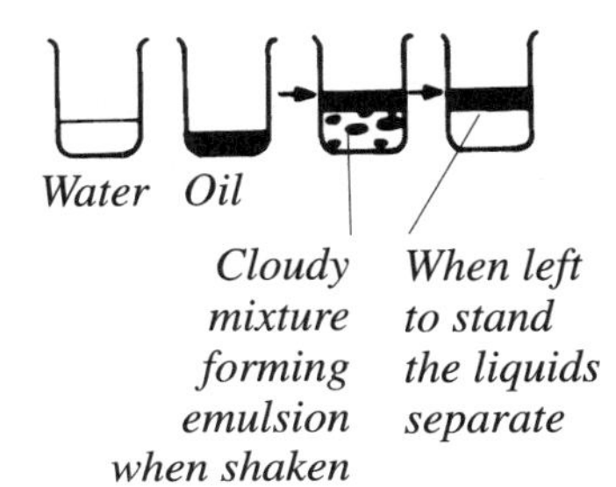

Emulsion

endothermic change A chemical reaction that absorbs heat from the surroundings.

end point The point at which a reaction is complete. The end point of a titration is the point at which one of the reactants has been completely used and it can be seen using an indicator.

energy change *See* exothermic change, endothermic change, enthalpy.

energy-level *See* electronic structure of atom.

energy-level diagrams These show the energy levels of reactants and products of a chemical reaction. They can be used to show changes in enthalpy.

enol A type of organic compound containing a hydroxyl group adjacent to a carbon atom that also has a double bond.

enthalpy A measure of the stored heat energy of a substance. During a chemical reaction, change in enthalpy can be measured. If energy is released (ΔH is negative), the reaction is exothermic; if energy is absorbed (ΔH is positive), the reaction is endothermic. $H = U + pV$, where H is the enthalpy, U the internal energy of the system, p its pressure, and V its volume.

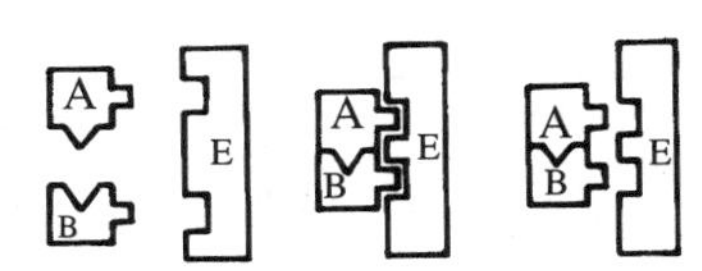

Enzyme action

enzyme An organic catalyst made of proteins. They are produced by living cells and increase the rate of a specific biochemical reaction. Enzymes can be manufactured using microorganisms or animals and plants, and are used in industrial processes.

Epsom salts *See* magnesium sulfate.

equation *See* balanced equation, ion–electron equation, ionic equation, word equation.

equilibrium The state of a reversible chemical reaction where the forward and backward reactions take place at the same rate (i.e. equilibrium is reached when there is no apparent change in the amounts of reactants and products with time).

equilibrium constant *See* law of mass action.

equivalent mass (equivalent mass or combining mass) The mass of an element or compound that will combine with or replace 8 g of oxygen (or 1 g of hydrogen or 35.5 g of chlorine) in a chemical reaction. For an element, it is the element's relative atomic mass divided by its valency.

equivalent weight of acid An alternative to using molarities in calculations of acid/alkali titrations. The mass (in grams) of an acid that can produce 1.008 g of hydrogen ions when dissolved in water is its equivalent weight.

equivalent weight of base The mass (in grams) of an alkali which reacts with the equivalent weight of an acid.

Er Symbol for the element erbium.

erbium Element symbol, Er; rare earth element/lanthanide; Z 68; A(r) 167.26; density (at 20°C), 9.07 g/cm^3; m.p., 1529°C; named for Ytterby, a town in Sweden; discovered 1843; oxide is used in glass manufacture.

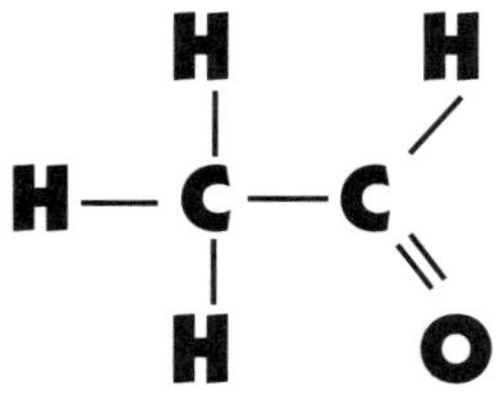

Es Symbol for the element einsteinium.

esterification The formation of an ester formed by the reaction of an organic acid and an alcohol.

esters A group of hydrocarbons that are formed by a reaction between a carboxylic acid and an alcohol. Esters are used in flavorings and perfumes because they have a sweet fruity smell.

Ethanal

ethanal CH_3CHO (acetaldehyde) A colorless liquid; b.p., 20.8°C. It is soluble in water.

ethane C_2H_6. An alkane. A colorless flammable saturated gas; m.p., –183°C; b.p., –89°C. It occurs in natural gas.

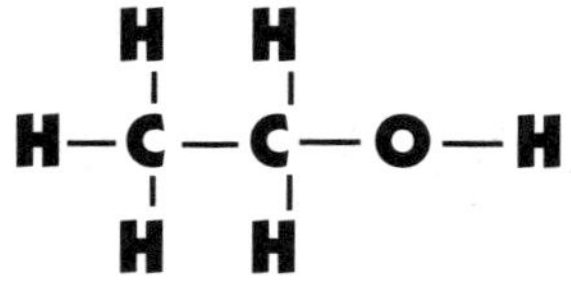

ethane-1,2-diol CH_2OHCH_2OH (ethylene glycol) A dihydric alcohol. A colorless, water-soluble, viscous hygroscopic liquid; m.p., –13°C; b.p., 197°C. It is used as antifreeze and in the manufacture of polyesters.

Ethanol

ethanoic acid CH_3 COOH (acetic acid) A weak organic acid. Its salts are ethanoates (acetates).

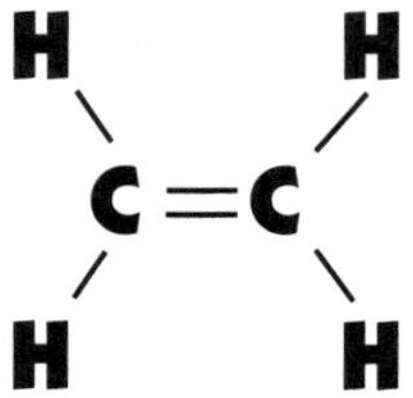

ethanol C_2H_5OH(ethylalchol) An alkanol. A volatile, colorless water-soluble liquid; m.p., –117°C; b.p., 78.5°C. It is manufactured by the fermentation of certain carbohydrates in the alcoholic drinks industry. Industrial ethanol is manufactured by the hydrolysis of ethane. It is used as a solvent.

ethanoyl group The organic group CH_3CO-.

Ethene

ethene C_2H_4 (ethylene). An alkene. A colorless flammable unsaturated gas; m.p., –169°C; b.p., –102°C. It is manufactured by cracking petroleum gas, and it is used in ethanol and poly(ethene) production.

ethers A group of organic compounds containing the group –O–. They are volatile liquids with a pleasant smell and they are insoluble in water but soluble in alcohol. Diethyl ether ($C_2H_5OC_2H_5$) is the simplest ether and is known as *ether*. It was formerly used as an anesthetic and is now used as a solvent.

ethoxyl group $-O.C_3H_5$.

ethyl acetate *See* ethyl ethanoate.

ethyl alcohol *See* ethanol.

ethylene *See* ethene.

ethylene dibromide *See* 1,2–dibromoethane.

ethylene glycol *See* ethane-1,2-diol.

ethyl ethanoate $CH_3COOC_2H_5$. A sweet-smelling ester produced by the reaction between ethanoic acid and ethanol. It is used in glues and paint as a solvent.

ethyl group or **radical** $C_2H_5^-$. It is present in many organic compounds.

ethyne C_2H_2 (acetylene) An alkyne. A colorless flammable unsaturated (contains a triple bond between the carbon atoms) gas; m.p., –82°C, b.p., –84°C. It was formerly made by the action of water on calcium carbide and is now made from cracking petroleum products. When combined with oxygen and burnt, high temperatures (3000°C) are reached; this mixture is used in the oxyacetylene torch for cutting and welding metals.

$$H-C\equiv C-H$$

Ethyne

Eu Symbol for the element europium.

europium Element symbol, Eu; rare earth element/lanthanide; Z 63; A(r) 151.96; density (at 20°C), 5.24 g/cm^3; m.p., 822°C; named for the continent of Europe; discovered 1896; used as neutron absorber in nuclear reactor and compounds used in color televisions.

eutectic alloy An alloy consisting of at least two metals in proportions that gives the lowest melting point of any composition of the metals.

eutectic mixture A mixture of two or more substances that melts at the lowest freezing point of any mixture of the components. This temperature is the eutectic point. The liquid melt has the same composition as the solid.

eutrophic Containing too many nutrients. If land has been overfertilized, water running over it will carry large amounts of nitrates and phosphates into rivers and lakes. These nutrients cause rapid growth of weeds, which choke the water, removing oxygen and preventing sunlight from penetrating to the lower levels of the water. This affects the ability of the river or lake to support animal and plant life and it is said to be suffering from eutrophication.

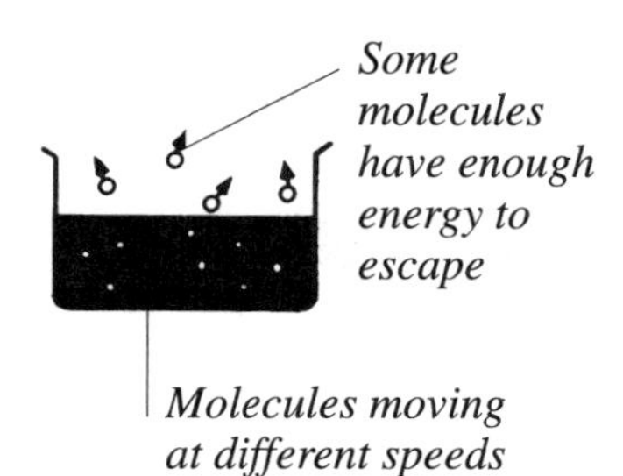

Evaporation

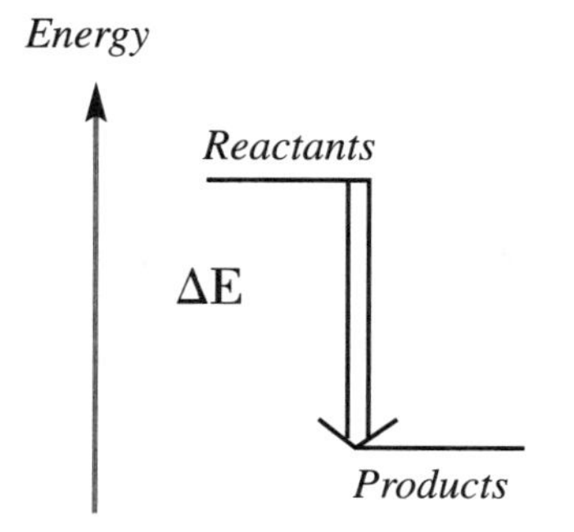

Exothermic

evaporation The process in which a liquid changes state to vapor. It can occur at any temperature up to the boiling point of the liquid. An evaporator is a system in which evaporation can take place.

exothermic changes A chemical reaction that releases heat to the surroundings.

F Symbol for the element fluorine.

Faraday constant The amount of electricity needed to liberate one mole of a monovalent ion during electrolysis ($9.648\ 670 \times 10^{-4}$ C mol^{-1}).

Faraday's laws of electrolysis (1) The amount of chemical change (or mass of a substance liberated at an electrode) produced by a current is proportional to the quantity of electricity passed.

(2) The quantities of different substances deposited or liberated by a given quantity of electricity are in the ratio of their chemical equivalent weights.

fat (lipid) The general name for mixtures of triglycerides (*see* glycerides) of fatty acids. They have a melting point above room temperature.

fatty acid (alkanoic acid) The general formula is $C_nH_{2n+1}COOH$. A fatty acid is a straight chain saturated or unsaturated monocarboxylic acid (having one carboxyl group). The higher (longer-chain) fatty acids occur in nature and combine with glycerol to form esters such as oils and fats.

Fe Symbol for the element iron.

Fehling's solution An aqueous solution prepared by mixing copper(II) sulfate ($CuSO_4$) solution with an alkaline solution of potassium sodium tartarate sodium hydroxide. It is used to test for estimating and detecting reducing sugars.

Fehling's test If a mixture of an aqueous solution of a reducing sugar and Fehling's solution is boiled, a brick red precipitate of copper (I) oxide confirms that the solution contained a reducing sugar.

fermium Element symbol, Fm; actinide; Z 100; A(r) 257; very radioactive; named in honor of Enrico Fermi; discovered 1952.

ferroxyl indicator A pale yellow solution that turns blue in the presence of Fe^{2+} ions. It is used to test for rust.

fertilizers Natural (farmyard manure, compost) or synthetic substances (such as ammonium nitrate and superphosphate) added to the soil to replace nutrients used by crops. Synthetic fertilizers are manufactured to contain the elements nitrogen, phosphorus, and potassium, which are required by plants.

film Any thin layer of a substance.

filter A device, containing a porous material such as paper or sand, that removes suspended solid particles from a fluid.

filtrate Clear liquid that has passed through a filter.

fire extinguisher Carbon dioxide is used in fire extinguishers as it does not support burning and therefore smothers a fire. Some fire extinguishers contain an acid and sodium carbonate (or sodium hydrogen carbonate), and when these reactants mix, carbon dioxide is produced. Alternatively, a fire extinguisher can contain pressurized carbon dioxide.

fire-damp Common name for methane found in coal mines.

fission A process (spontaneous or induced) during which a heavy atomic nucleus disintegrates into two lighter atoms that together have less mass than the total initial material. This lost mass is converted to energy, the amount of which is given by Einstein's equation $E = mc^2$.

flame A burning mass of gas that gives out heat and light energy.

flame test This test allows salts containing metal ions to be identified, due to the observation that the presence of certain metal ions causes a coloration in a flame in which these ions are burnt.

In order to test a solution, a clean nichrome or platinum wire is bent and cleaned. This is done by dipping it into concentrated hydrochloric acid, putting it in the hottest part of a Bunsen flame until it no longer colors the flame, and dipping it into concentrated hydrochloric acid again. The wire is then dipped into a sample of the solid to be tested and placed in the flame.

If the flame burns yellow/orange, the sample contains sodium ions; lilac denotes potassium ions; brick red denotes calcium; apple green denotes barium; green/blue denotes copper; bright red denotes lithium; red denotes strontium.

flocculation The grouping together of colloidal particles to form a precipitate that may float in the liquid.

fluid A substance that can flow because its particles are not fixed in position. Liquids and gases are fluids.

fluorescence The emission of light from an object that has been irradiated by light or other radiations. Energy is absorbed by the object and then re-radiated at a longer wavelength than the incident light.

fluoridation Very small amounts of fluorides (1 part per million of fluoride ions) added to drinking water to prevent dental disease.

fluorine Element symbol, F; halogen, group 7; very reactive gas; Z 9; A(r) 19; density (at 20°C), 1.696 g/l at STP; m.p., –219.6°C; name derived from the Latin *fluere*, also fluorspar; discovered 1771; used in manufacture of plastics and toothpaste and in water treatment.

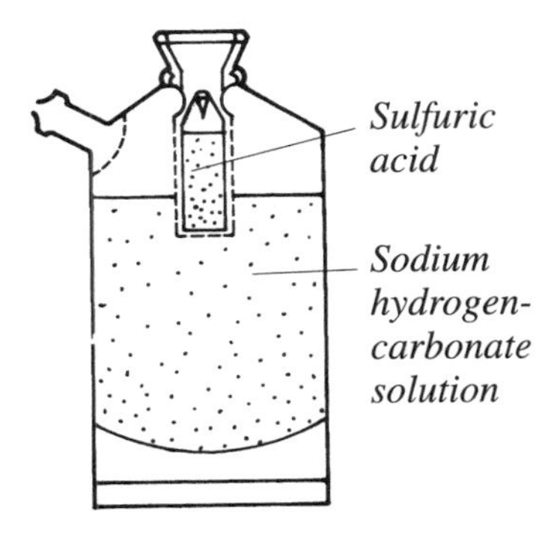

Fire extinguisher

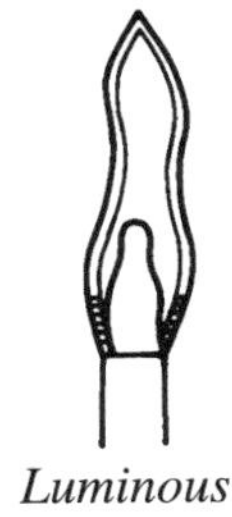

Types of flame

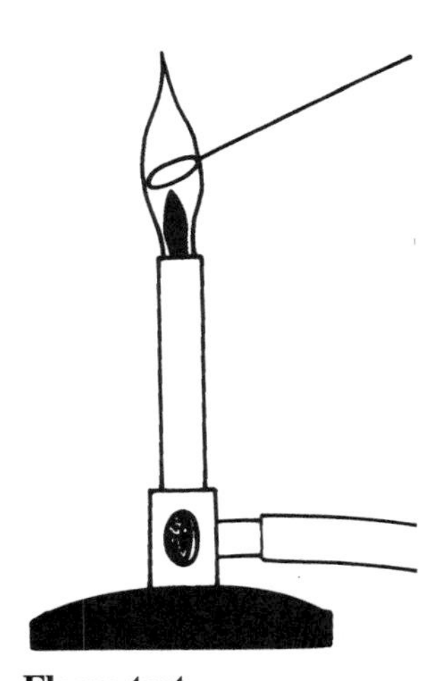

Flame test

flux A substance that combines with another substance (usually an oxide) forming a compound with a lower melting point than the oxide.

Fm Symbol for the element fermium.

foam A dispersion of gas in a liquid or solid. Small bubbles of gas are separated by thin films of the liquid or solid.

f orbital A type of orbital, of which seven types are possible, each of which can hold two electrons.

formaldehyde *See* methanal.

formalin A solution of methanal in water. It is used as a preservative.

formic acid *See* methanoic acid.

formula (often used to refer to the molecular formula) A way in which the composition of a chemical compound may be represented using symbols to represent the atoms present. *See* empirical formula, full structural formula, general formula, molecular formula, perspective formula, shortened structural formula, simple formula.

formula mass The relative molecular mass of a compound calculated using its molecular formula. It is also the mass of a mole of the substance.

fossil fuel Coal, oil, and natural gas are fossil fuels. Coal was formed about two hundred million years ago by the bacterial decomposition of plants, such as large tree ferns and giant reeds, followed by exposure to heat and pressure. Oil and natural gas were formed in the oceans millions of years ago from microscopic plants and animals that sank to the bottom of the sea. They were covered with layers of sand and other materials that subjected them to pressure and helped to turn them into oil and gas.

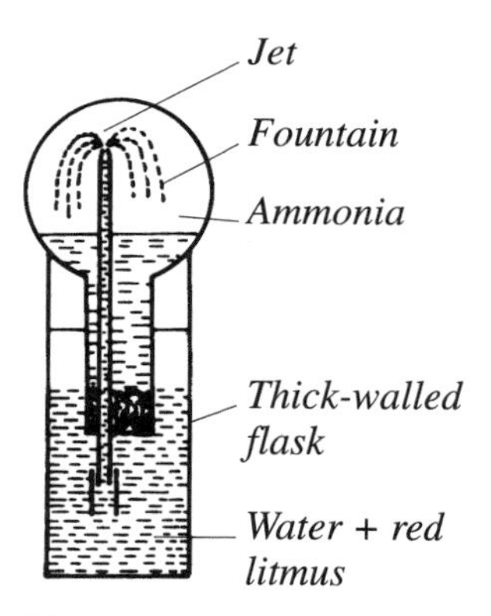

Fountain experiment

fountain experiment Demonstration of the solubility of ammonia. If a large dry flask full of ammonia, closed with a stopper and containing a glass tube, is inverted in a tall jar of water colored with red litmus solution, the ammonia in the glass tube dissolves and the water rises up the tube. When a drop of water reaches the flask, it dissolves most of the ammonia, leaving a partial vacuum in the flask. Water from the jar is forced up the glass tube by atmospheric pressure, fountaining into the flask. The water turns blue on entering the flask because of the presence of the dissolved ammonia.

Fr Symbol for the element francium.

fractional distillation The separation of a mixture of liquids that have differing but similar boiling points. The fractionating column allows the separate liquids (or fractions) in a mixture to be collected at different temperatures. The temperature is higher in the lower regions of the fractionating column, which is where the less volatile compounds condense and are removed. The more volatile compounds progress up the column to condense at lower temperatures.

fractional distillation of oil The use of fractional distillation to separate oil into different fractions.

The least volatile compound (the residue with a boiling point above 400°C) is bitumen; it is used for road surfaces and roofing material.

Heavy oil is collected in the boiling range 350–400°C; it is used for lubricating oil, fuel oil for furnaces, Vaseline, and paraffin wax.

Gas oil and diesel oil are collected in the boiling range 250–350°C and they are used as fuel.

Kerosene (paraffin oil or naphtha) is collected in the boiling range 175–250°C. This fraction is cracked to form gasoline.

Gasoline is collected in the boiling range 50–175°C. This is used as motor fuel.

Hydrocarbon gases are collected at temperatures below 40°C. They are used as "bottled gas."

fractionating column *See* fractional distillation.

francium Element symbol, Fr; alkali metal group 1; radioactive (half-life of most stable isotope is 21 minutes), occurs in the radioactive series; Z 87; A(r) 223; density (at 20°C), 2.4 g/cm^3; m.p., 27°C; named for France; discovered 1939.

Frasch process The process in which sulfur is extracted from deep underground deposits. It consists of three concentric pipes that are sunk to the level of the deposit. Superheated water is forced down the outermost pipe and hot compressed air through the innermost pipe. As the steam melts the sulfur, it is forced up the middle pipe with air and water. Sulfur solidifies in large tanks on the surface.

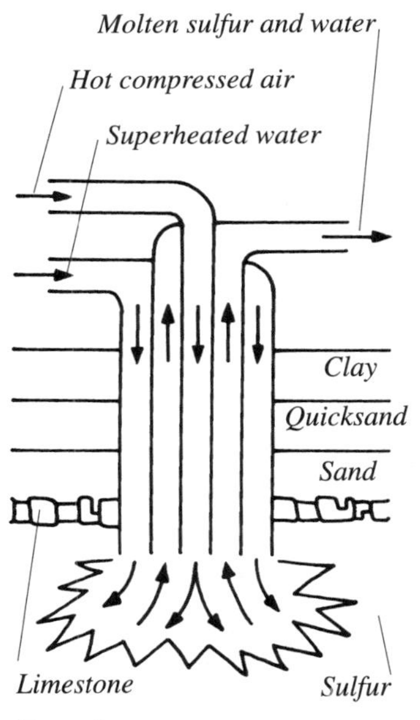

Frasch process

freeze-drying The removal of water from a frozen substance by reducing the pressure and allowing the water to sublime. This process is used for the dehydrating of heat-sensitive substances such as blood plasma and food.

freezing The process by which a change of state from liquid to solid occurs. The freezing point is the temperature at which this change occurs (it is also the temperature of the melting point, when the state changes from solid to liquid). It is the point at which the solid and liquid are in equilibrium. A freezing mixture is used to create a low temperature for chemical reactions. The mixture absorbs heat, producing lower temperatures than the original components of the mixture.

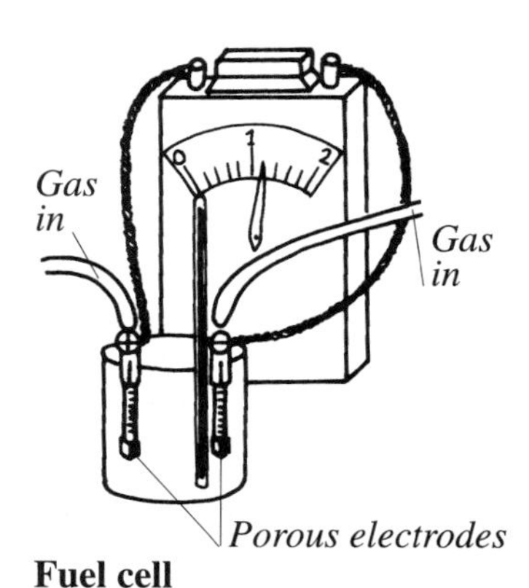

Fuel cell

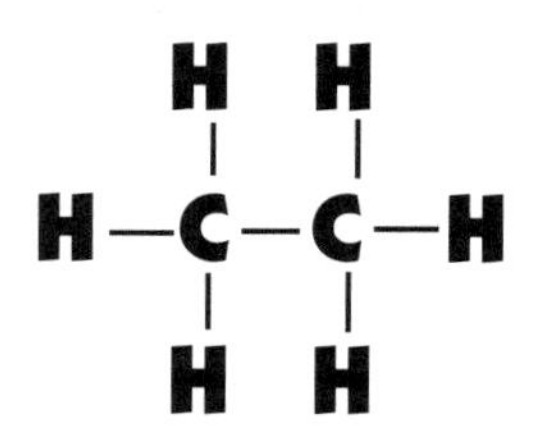
Full structural formula

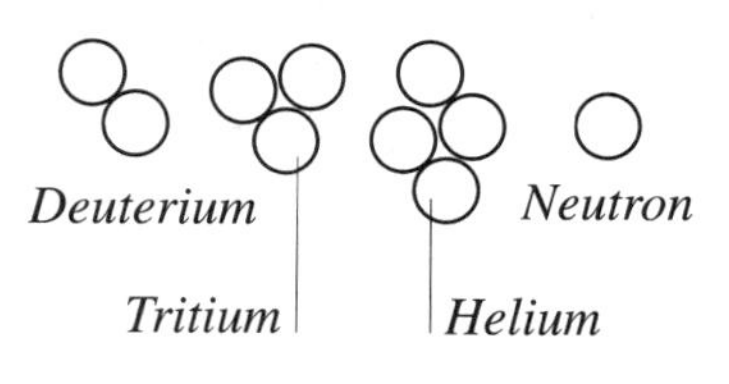

Fusion reaction

Freon *See* chlorofluorocarbons.

fructose $C_6H_{12}O_6$ (fruit sugar). The sweetest sugar. It occurs in fruits and honey.

fuel A substance that can produce large quantities of heat by either burning or undergoing nuclear fission.

fuel cell A type of primary cell that operates in the opposite way to electrolysis. It converts chemical energy directly to electrical energy. One type of fuel cell passes hydrogen and oxygen over porous electrodes where electricity is produced and the gases are converted to water.

full structural formula This represents the atoms in a molecule and the bonds between them, giving an indication of their position in relation to each other, although it does not always show the actual positions.

functional group (organic chemistry). The atom (or group of atoms) present in a molecule, which determines the characteristic properties of that molecule.

fundamental particles The large number of subatomic particles making up the universe. Many subatomic particles have been discovered and they are of three main types:

(1) leptons: particles of low mass, such as electrons, muons, tauons and associated neutrinos;

(2) mesons: unstable particles of medium mass consisting of two quarks;

(3) baryons: more massive particles consisting of three quarks. The proton and neutron are baryons.

Each particle has an anti-particle (a particle with the same mass but opposite charge). The fundamental particles in the nucleus are quarks.

fundamental units Internationally agreed, independently defined units of measurement used to form the basis of a system of units. As a base for such a system, three mechanical units, such as length, mass, and time, and one electrical unit are required. SI units (using the meter, kilogram, and second, together with the kelvin, candela, and mole) are the standard. Formerly, the c.g.s. (using the centimeter, gram and second) and the m.k.s. (meter, kilogram, second) and the f.p.s. (foot, pound, second) systems have been used (*see* SI).

fusion **1.** (*melting*) The process by which a change of state from solid to liquid occurs.

2. (*nuclear*) The process (which requires extremely high temperatures to initiate) by which two or more light atomic nuclei join, forming a single heavier nucleus. The products of fusion are lighter than the components. The mass lost is liberated as energy, given by Einstein's equation $E = mc^2$.

Ga Symbol for the element gallium.

gadolinium Element symbol, Gd; rare earth element/lanthanide; Z 64; A(r) 157.25; density (at 20°C), 7.9 g/cm^3; m.p., 1313°C; named in honor of Gadolin, a Finnish chemist; discovered 1880; used in electronics industry; compounds used as catalysts.

galena Metallic looking, naturally occurring ore consisting of PbS.

gallium Element symbol, Ga; Group 3; Z 31; A(r) 69.74; density (at 20°C), 5.9 g/cm^3; m.p., 29.8°C; name derived from the Latin name for France, *Gallia*; discovered 1875; used in semiconductors.

galvanizing The coating of iron or steel plates with a layer of zinc to protect against rusting. It is done either by dipping the iron or steel into a bath of molten zinc or by electrolysis.

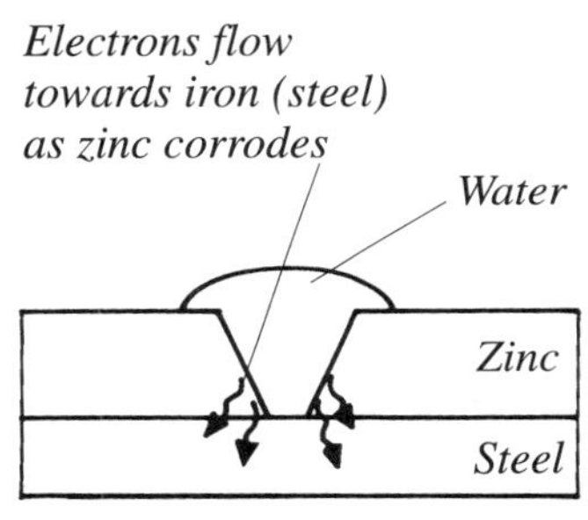

Galvanizing

gamma radiation Very short-wave electromagnetic radiation emitted as a result of radioactive decay. It is the least ionizing and most penetrating of the three types of radiation emitted in radioactive decay. It will penetrate a thick metal sheet and is only stopped by over 15 cm of lead or by thick layers of concrete.

gas One of the three states of matter. In a gas, the particles can move freely throughout the space in which it is contained. Gas is the least dense of the states of matter.

gas law The equation combining Boyle's law and Charles' law, $PV = nRT$, where P is the pressure, V is the volume, n is the number of moles of gas present, T is the temperature measure in Kelvin, and R is the universal gas constant.

gasoline A mixture of alkanes with chain lengths of between five and ten carbon atoms used as a fuel for internal combustion engines. It is obtained from the fractional distillation of petroleum and from cracking and reforming of hydrocarbons. *See* octane rating.

Gay-Lussac's law Volumes of gases that react do so in simple whole number ratios to each other and to the volumes of any gaseous products. (The volumes are measured at constant temperature and pressure.)

Gd Symbol for the element gadolinium.

Ge Symbol for the element germanium.

gel A colloidal solution that has formed a jelly. The solid particles are arranged as a fine network in the liquid phase.

general formula A formula showing the relative numbers of atoms present using the variable "n" for members of a homologous series.

germanium Element symbol, Ge; group 4; grayish white metalloid; Z 32; A(r) 72.59; density (at 20°C), 5.36 g/cm^3; m.p., 937.4°C; has excellent semiconductor properties; name derived from the Latin name for Germany, *Germania*; discovered 1886; used in manufacture of electronic components.

getter A substance that is evaporated on the inside surface of a vacuum tube to adsorb residual gas.

giant structure Atoms or ions present in very large numbers in a lattice. Each particle has a strong attractive force for those around it; this spreads the effect of the forces through the structure. Ionic compounds have giant structures, as do most elements (all metals and several nonmetals). Giant structures have high melting and boiling points. *See* covalent network.

Gibbs function or **Gibbs free energy** The energy absorbed or released in a reversible reaction at a constant temperature or pressure. It is calculated for a system from the enthalpy minus the product of the entropy and absolute temperature.

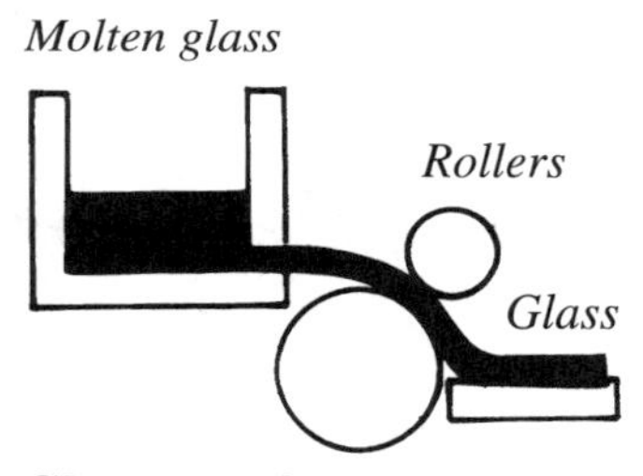

Glass manufacture

glass or **soda glass** A transparent substance formed by the fusion of silicon dioxide (white sand) with carbonates or oxides of calcium, sodium, potassium, or lead. Hard glass is a mixture of potassium and calcium silicates.

glucose $C_6H_{12}O_6$ (grape sugar, blood sugar). It is found in fruit juices, plant leaves, and in animal blood, and it is formed in plants by photosynthesis.

$6CO_2 + 6H_2O$ + energy (chlorophyll catalyst) = $C_6H_{12}O_6 + 6O_2$.

The enzyme zymase in yeast causes glucose to ferment to ethanol.

glycerides These are esters that are formed between glycerol and one or more organic acids. Depending on the number of hydroxyl groups that have reacted with the fatty acids, the glyceride may be a mono-, di-, or triglyceride. Glycerides made from unsaturated fatty acids usually have lower melting points than those made with saturated fatty acids, which have the same number of carbon atoms.

glycerol (propane-1,2,3-triol) $CH_2OH–CHOH–CH_2OH$. A trihydric alcohol. It is a colorless, water-soluble, viscous, hygroscopic liquid.

glycol *See* ethane-1,2-diol.

gold Element symbol, Au; transition element; shiny yellow metal; Z 79; A(r) 196.97; density (at 20°C), 19.32 g/cm^3; m.p., 1064.4°C; good conductor of heat and electricity; Old English name, *gold*, Latin *aurum*; known since prehistoric times; used in coins, jewelry, and electrical contacts.

Graham's law The velocity with which a gas will diffuse is inversely proportional to the square root of its density.

gram-equivalent mass The equivalent mass expressed in grams. It is the number of grams of the element (or compound) that combines with or replaces 8 g of oxygen (or 35.5 g of chlorine, or 1 g of hydrogen) during a chemical reaction.

gram-formula mass An alternative way of describing a mole of a substance.

graphite An allotrope of carbon; m.p., 3650°C. It is an opaque, grayish black solid with a metallic luster. It has a structure of hexagonal crystals arranged in giant layers that slide over each other because the forces between the layers are weak. It is a good conductor of heat and electricity.

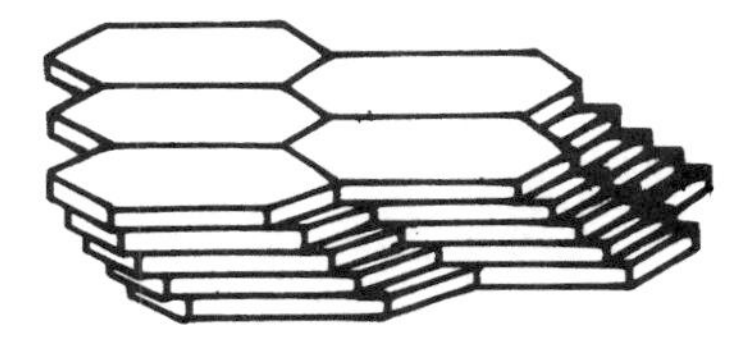

Graphite sliding

gravimetric analysis A method of quantitative analysis in which accurate results are obtained by determining the weights of the components of a compound. For example, if the compound whose components were to be measured was in solution, an insoluble salt of one of the components could be precipitated and then weighed.

greenhouse effect Atmospheric warming caused by gases that act like a greenhouse roof, trapping solar heat below them. Waste gases produced by human industrial and agricultural activity are arguably intensifying the natural greenhouse effect.

Greenhouse effect

green vitriol Hydrated iron(II) sulfate. $FeSO_4.7H_2O$ (iron sulfate heptahydrate). It is also known as copperas. *See* iron sulfates.

ground state The lowest energy state of an atom in which the electrons occupy the orbitals of lowest available energy. If an atom has more energy than it would possess in the ground state, it is said to be in an excited state.

group The vertical columns of elements in the periodic table. Elements in a group react in a similar way. There is a gradation in properties from one member of the group to the next. They have the same number of electrons in their outer shell and an increasing number of shells. Elements lower in a group have increased metallic character.

group 1 or **I elements** The alkali metals. The elements lithium, sodium, potassium, rubidium, cesium, francium. These elements have one electron in their outer shell. They are very electropositive, soft, less dense than water, and have low melting points. They form strong alkalis and have a valency of one.

group 2 or **II elements** The alkaline earth metals. The elements beryllium, magnesium, calcium, strontium, barium, radium. These elements have two electrons in their outer shell. They are electropositive and are harder and less dense than the group 1 elements.

group 3 or **III elements** The elements boron, aluminum, gallium, indium, and thallium. These elements have a full s orbital and one electron in a p orbital in their outer shell. Their inner shells are full. There is more variation in properties in this group than in groups 1 and 2. Elements lower in the group are more metallic than those higher in the group and are more likely to form ions with a positive charge of 3.

group 4 or **IV elements** The elements carbon, silicon, germanium, tin, and lead. These elements have a full s orbital and two electrons in two p orbitals in their outer shell. Their inner shells are full. The character of the group changes from nonmetallic at the top to metallic at the bottom. The elements have a valency of 4, although the larger atoms (lower in the group) tend to form divalent compounds.

group 5 or **V elements** The elements nitrogen, phosphorus, arsenic, antimony, bismuth. These elements have a full s orbital and three electrons in three p orbitals in their outer shell. Their inner shells are full. The two lightest elements are nonmetals and the others are metalloids. Nitrogen has a valency of 3 and forms covalent compounds. The character of the group increases in its metallic nature further down the group. The elements in this group form compounds with the 3 oxidation state. The larger elements can also form compounds with the 5 oxidation state by using the available d orbitals in their outer shell (they promote an s electron from the outer shell to a d orbital).

group 6 or **VI elements** (the chalcogens) The elements oxygen, sulfur, selenium, tellurium, and polonium. These elements have a full s orbital, one full p orbital, and two half-full p orbitals in their outer shell. Their inner shells are full. As the outer shells of these elements are only two electrons short of the noble gas structure they tend to be electronegative (this tendency decreases in the larger elements) and nonmetallic. These elements form covalent bonds with a variety of other elements and they all form hydrides with two atoms of hydrogen.

group 7 or **VII elements** (the halogens) The elements fluorine, chlorine, bromine, iodine, astatine. These elements have a full s orbital, two full p orbitals, and one half-full p orbital in their outer shell. Their inner shells are full. As the outer shells of these elements are only one electron short of the noble gas structure, they tend to be very electronegative, having high electron affinities and forming compounds by gaining an electron to form a stable outer shell. They can also share their outer electrons to form covalent compounds with single bonds, and all exist as diatomic molecules.

group 8 or **VIII elements** (the noble or inert gases) The elements helium, neon, argon, krypton, xenon, radon. The outer shell of the atoms in these elements is complete, rendering these elements unreactive.

H Symbol for the element hydrogen.

Haber process This is used in the industrial manufacture of ammonia. Nitrogen and hydrogen are dried, mixed, and reacted together at high temperature and pressure in the presence of a catalyst to form ammonia. As only about 15% of the reactants combine under typical conditions, the unreacted nitrogen and hydrogen are recycled for further reaction.

hafnium Element symbol, Hf; transition element; shiny silvery metal; Z 72; A(r) 178.49; density (at 20°C), 13.3 g/cm^3; m.p., 2227°C; name derived from *Hafnia*, the Latin name for Copenhagen; discovered 1923; used in alloy with tungsten to make filaments and electrodes. Used in control rods of nuclear reactors to absorb neutrons.

half-cell A metal in contact with a solution of its own ions. If two half-cells (each using a different metal) are connected together, electricity is produced. To do this, the metals are connected by wires and the solutions in the cells are joined by an ion bridge. A strip of filter paper soaked in a solution of sodium chloride can form an ion bridge. The wires between the metal rods and the ion bridge connecting the solutions completes the circuit.

half-life A substance that undergoes exponential decay decays by the same ratio in equal intervals of time. The constant ratio is the half-life. The rate of radioactive decay of a substance is defined by its half-life.

half-reaction A representation of a reaction that is particularly useful when considering redox reactions. A complete reaction between two substances is viewed as two separate reactions of each of the substances. For example, the displacement of zinc ions from solution by magnesium

$Zn^{2+}(aq) + Mg(s) \rightarrow Mg^{2+} + Zn(s)$.

This may be seen as $Zn^{2+}(aq) + 2e^- \rightarrow Zn(s)$

and $Mg(s) \rightarrow Mg^{2+} + 2e^-$

halide A compound that a halogen makes with another element. Metal halides are ionic; nonmetal halides are formed by covalent bonding.

halogen *See* group 7 elements.

halogenation The introduction of one or more halogen atoms into the structure of an organic molecule. If the halogen introduced is chlorine, the process is termed chlorination; if the halogen is bromine, the process is bromination, etc.

hardness *See* Mohs' scale.

hardness in water The presence of calcium and magnesium ions in water, which restricts the ability of soap to form a lather and leaves deposits

Nitrogen from air

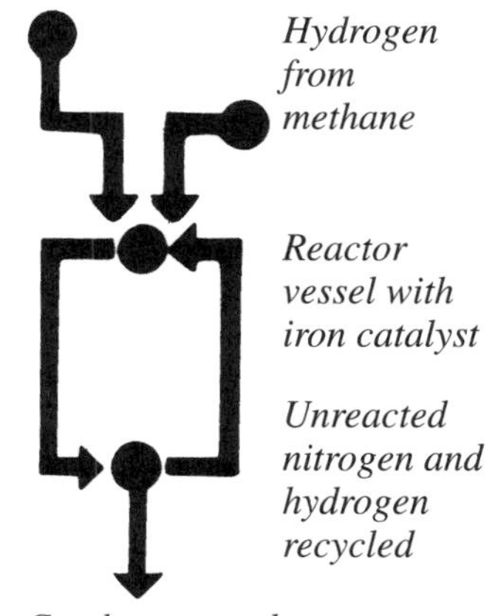

Condenser produces liquid ammonia

Haber process

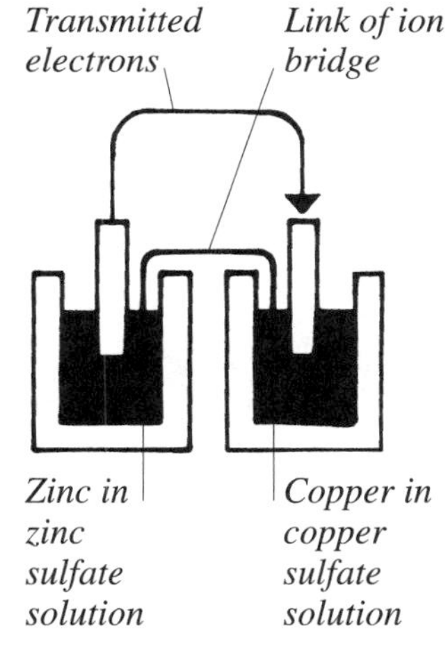

Half-cells

$CH_4 + Cl_2 \rightarrow CH_3Cl + HCl$

$C_6H_6 + Br_2 \rightarrow C_6H_5Br + HBr$

Halogenation

in pipes. There are two types. In temporary hardness, soluble hydrogencarbonate compounds of calcium, magnesium, and iron are dissolved in cold water. When the water is heated, these salts form insoluble carbonates that are precipitated. Permanent hardness is caused largely by calcium sulfate and cannot be easily removed.

hassium Element symbol, Hs; transition element; Z 108; A(r) 265; name derived from the Latin name for Hess (the German state), *Hassias*; discovered 1984. Formerly known as unniloctium.

He Symbol for the element helium.

heat capacity The heat capacity of an object is the product of its mass and its specific heat capacity.

heat energy A system has heat energy because of the kinetic energy of its atoms and molecules (due to translation, rotation, and vibration). It is transferred by conduction, convection, and radiation.

heat exchange A process of transferring energy from one material to another. If a process is well designed, the heat produced in one part of a system can be used in another.

heat of combustion The heat change when one mole of a substance is completely burnt in oxygen (both products and reactants being at 25°C and one atmosphere). The value is negative when heat is given out (the change is exothermic).

heat of formation The heat change when one mole of a compound is formed from its elements at 25°C and one atmosphere.

heat of neutralization The heat change when acid that produces one mole of hydrogen ions is neutralized by an alkali (both acid and alkali being in dilute solution). The heat of neutralization of a strong acid by a strong alkali.
The reaction $H^+(aq) + OH^-(aq) \rightarrow H_2O(l)$ is usually about –57 kJ.

heat of reaction The difference between the enthalpy of the products of a reaction and the enthalpy of the reactants. (The heat of reaction is negative if the reaction is exothermic and positive if the reaction is endothermic.)

heat of solution The heat change when one mole of solute dissolves in a large volume of solvent (usually water) until no further heat change is observed.

heavy metals High-density metals such as cadmium, lead, and mercury. Heavy metals are poisonous, and careless dumping of heavy-metal wastes can create local health hazards.

heavy water D_2O. Water containing two atoms of deuterium (the isotope of hydrogen whose relative atomic mass is twice that of hydrogen) in place of hydrogen.

helium Element symbol, He; noble gas, group 8; Z 2; A(r) 4; density (at 20°C), 0.178 g/l at STP; m.p., –272.2°C; chemically inert; name derived from the Greek *helios*, "sun"; discovered 1895; used to fill balloons, as an inert atmosphere for arc welding, and in gas lasers.

hemoglobin A red oxygen-carrying pigment found in the red blood cells of vertebrates and in the blood plasma of some invertebrates.

Henry's law The mass of a gas that dissolves in a given volume of solvent at a constant temperature is proportional to the pressure of the gas (assuming that the gas does not react with the solvent).

Hess's law The total heat change during a complete chemical reaction is the same regardless of the number of intermediate stages that take place in the reaction.

heterocyclic compound An aromatic organic compound where one or more atoms other than carbon form part of the ring structure.

heterogeneous catalyst The catalysis of a reaction by a substance that is in a different state (solid, liquid, or gas) from the reactants.

heterogeneous reaction A chemical reaction taking place between substances in different physical states – solids, liquids, and/or gases.

Hf Symbol for the element hafnium.

Hg Symbol for the element mercury.

Ho Symbol for the element holmium.

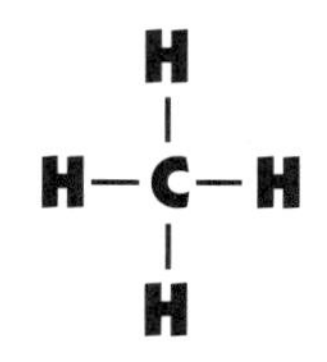

holmium Element symbol, Ho; rare earth/lanthanide; Z 67; A(r) 164.93; density (at 20°C), 8.8 g/cm^3; m.p., 1474°C; name derived from *Holmia*, the Latin name for Stockholm; discovered 1879.

homogeneous catalyst The catalysis of a reaction by a substance that is in the same state (solid, liquid, or gas) as the reactants.

homogeneous reaction A chemical reaction taking place between substances that are in the same physical state – solid, liquid, or gas.

homologous series A series of related organic compounds. The formula of each member differs from the preceding member by the addition of a $-CH_2-$ group. Each series has a general formula; for example, the general formula for alkanes is C_nH_{2n+2}. The properties of each series, though similar, change gradually and regularly with increasing molecular weight.

Hs Symbol for the element hassium.

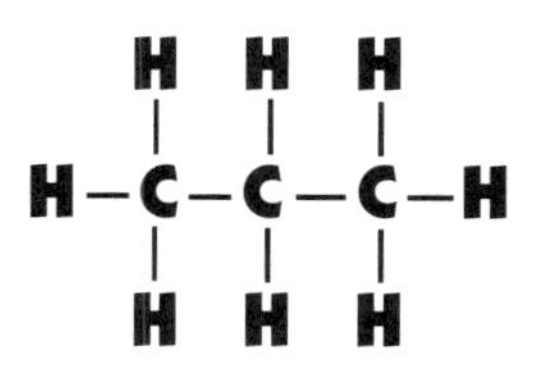

Homologous series

humidity The measure of the amount of water vapor in air, expressed as either the absolute or relative humidity.

Hund's principle When electrons are filling the orbitals of one type in a shell (for example the three p orbitals), they tend to distribute themselves

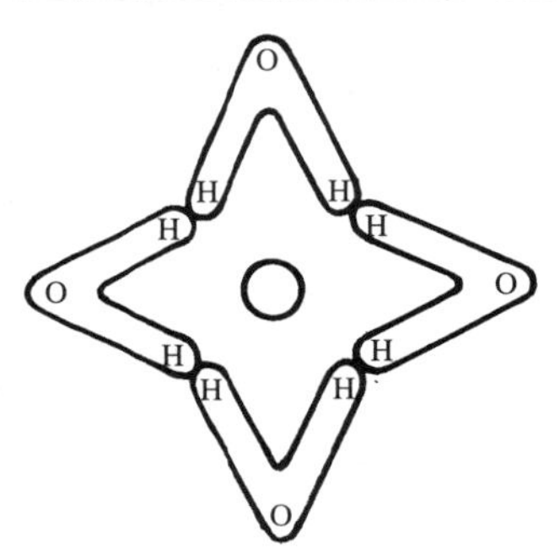

Anions surrounded by water molecules

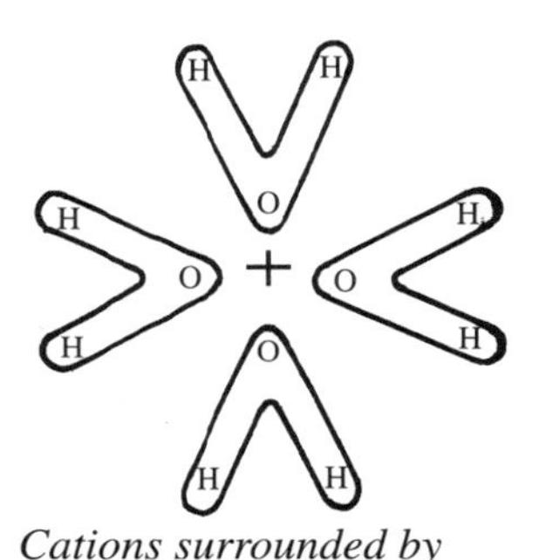

Cations surrounded by water molecules

Hydration

in different orbitals because they repel each other (if an atom had three p electrons they would each be in a different p orbital).

hydrate A salt containing water of crystallization, for example $CuSO_4.5H_2O$.

hydration A type of solvation. Molecules or ions of solute become surrounded by water molecules to which they are attached by coordinate bonds or by a type of electrostatic force.

hydride A compound formed between hydrogen and one other element. Electropositive elements form salt-like hydrides containing the hydride ion (H^-), which are very reactive. nonmetals and transition metals form covalent hydrides such as methane (CH_4), ammonia (NH_3), and water (H_2O).

hydriodic acid Aqueous solution of hydrogen iodide. A strong acid.

hydrobromic acid Aqueous solution of hydrogen bromide. It is a strong acid.

hydrocarbon chain A line of carbon atoms in a molecule. In polymers these chains can be thousands of atoms long. Molecules can consist of a straight chain or a branched chain.

hydrocarbon An organic compound that contains carbon and hydrogen only. Hydrocarbons can be aliphatic or aromatic and be saturated or unsaturated.

hydrochloric acid HCl. A strong acid. In dilute solution it reacts with most metals to form hydrogen. With carbonates it forms carbon dioxide and with sulfites it forms sulfur dioxide. With alkalis and insoluble bases it forms a salt and water. It contains hydrogen ions. The concentrated acid is a strong reducing agent.

hydrofluoric acid An aqueous solution of hydrogen fluoride. It is a weak acid that attacks glass and is used for glass etching.

hydrogen Element symbol, H; the lightest element; colorless gas; Z 1; A(r) 1.01; density (at 20°C), 0.09 g/l at STP; m.p., –259.1°C; explodes readily in oxygen; name derived from the Greek words *hydor*, "water," and *genes*, "producing"; discovered 1766; used in the synthesis of ammonia and the hydrogenation of oils.

hydrogenation The addition of hydrogen to another compound, usually an unsaturated organic compound. Nickel is a good catalyst for such reactions. Ethane is formed if ethene is hydrogenated. Hydrogenation is a very important process in the formation of margarine, in which unsaturated oils are hydrogenated to form saturated fats. The hydrogenation of vegetable and animal oils was first carried out in 1910, when the oils were heated to about 200°C and hydrogen was bubbled through them in the presence of finely divided nickel. The oils changed to fats which could be converted to margarine and other products.

hydrogen—atomicity of The hydrogen molecule is diatomic. *See* atomicity.

hydrogen bond This occurs in compounds in which a hydrogen atom makes a covalent bond with an electronegative element, for example, in water, H_2O. The bond is polarized because the electrons are attracted towards the electronegative oxygen atom, leaving the hydrogen atom with a positive charge.

If another water molecule approaches this hydrogen atom, the oxygen atom of the second water molecule forms a weak electrostatic bond, called a hydrogen bond, with the hydrogen atom in the first water molecule.

Hydrogen bonds are weaker than covalent or ionic bonds, but they do affect the physical properties of compounds in which they occur. *See* polar molecule.

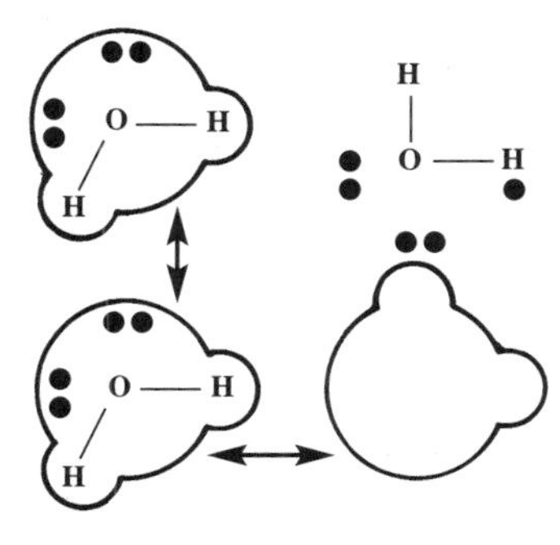

Electron pair

Hydrogen bond

hydrogen bromide HBr. A colorless gas; m.p., –86°C; b.p., –66.4°C.

hydrogencarbonates Acid salts of carbonic acid containing the ion $^{-}HCO_3$. *See* baking powder for sodium hydrogen carbonate. *See* hardness in water for other hydrogencarbonates.

hydrogen chloride HCl. A colorless gas with a pungent smell. It fumes in moist air, forming tiny drops of hydrochloric acid solution; m.p., –114°C; b.p., –85°C. It is very soluble in water.

hydrogen fluoride HF. A colorless liquid; m.p., –83°C; b.p., 19.5°C. Hydrogen bonds are formed in the liquid state. It is very corrosive (it is used in glass etching) and is a good fluorinating agent (*see* halogenation).

hydrogen iodide HI. A colorless gas (m.p., –51°C; b.p., –36°C) that is very soluble in water.

hydrogen ion H^+. This is usually regarded as being a single proton. In aqueous solution, the hydrogen ion exists in a hydrated form such as the oxonium (hydroxonium) ion H_3O^+.

hydrogen—isotopes of Hydrogen has three isotopes: the normal hydrogen atom (protium), whose nucleus contains one proton and thus has a relative atomic mass of one; deuterium, whose nucleus contains one proton and one neutron and thus has a relative atomic mass of two; tritium, whose nucleus contains one proton and two neutrons and thus has a relative atomic mass of three.

hydrogen peroxide H_2O_2. A colorless or pale blue viscous liquid; m.p., –0.89°C; b.p., 151.4°C. It decomposes in light to form water and oxygen. It is available in solutions designated by the volume of oxygen that can be liberated. For example, 20-volume hydrogen peroxide yields 20 volumes of oxygen from one volume of solution. It is a strong oxidizing agent, but it can also act as a reducing agent, depending on the substance with which it reacts. With lead(II) sulfide (PbS), it acts as an oxidizing agent, forming lead(II) sulfate ($PbSO_4$) and water. With lead(IV) oxide (PbO_2) it acts as a reducing agent, forming lead(II) oxide (PbO) oxygen and water).

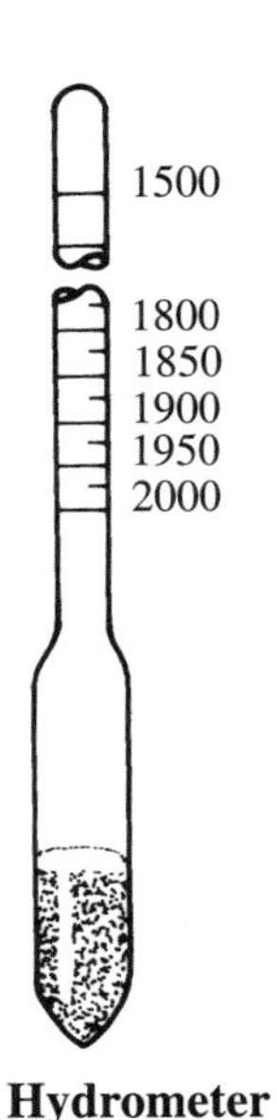

Hydrometer

hydrogen sulfide H_2S. A colorless poisonous gas smelling of bad eggs; m.p., –85.5°C; b.p., –60.7°C. It is moderately soluble in water; the solution is a weak acid (hydrosulfuric acid). Hydrogen sulfide is a reducing agent.

hydrolysis The interaction of water with a salt to form an acid and a base. The water dissociates to H^+ and OH^- ions.

hydrometer An instrument used to measure the densities or relative densities of liquids.

hydrophilic Water-loving. In solution, it refers to a chemical or part of a chemical that is highly attracted to water.

hydrophobic Water-hating. It refers to a chemical or part of a chemical that repels water.

hydroxide ion (hydroxyl ion or deprotonated water) The negative ion (OH^-) present in alkalis. It forms water with the addition of a hydrogen ion (H^+) or proton.

hydroxides A compound containing the hydroxide ion or the hydroxyl group bonded to a metal atom. Metal hydroxides are bases.

hydroxonium ion (hydronium ion or protonated water) The positive ion $(H_3O)^+$. It is the hydrated form of the hydrogen ion (H^+) or proton.

hydroxy A compound that contains one or more hydroxyl (OH^-) group.

hydroxyl group –OH. A monovalent functional group that consists of an oxygen atom and a hydrogen atom. In metal hydroxides it exists as the hydroxide (or hydroxyl) ion. *See* hydroxide ion.

hypochlorous acid HClO (chloric(I) acid) A weak acid that acts as a strong oxidizing agent. It forms salts called chlorates. Chlorates are used as bleaches and in water purification.

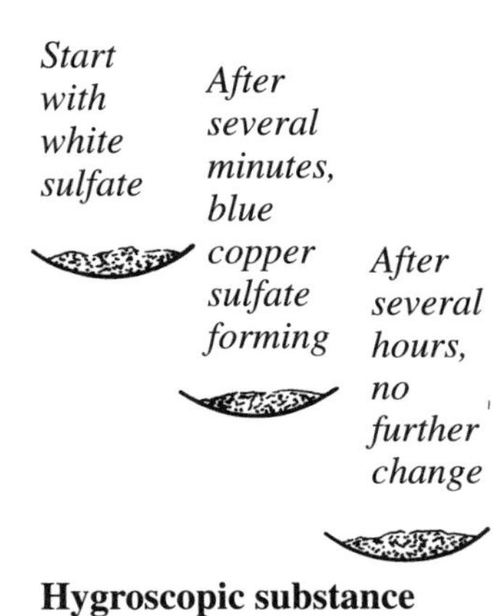

Hygroscopic substance

hygroscopic A substance that absorbs moisture from the air without becoming liquid.

I Symbol for the element iodine.

ice The solid state of water, H_2O. It is less dense than water and therefore floats on water.

ideal gas A gas that obeys the gas laws and in which molecules have negligible volume and the forces of molecular attraction are negligible. No known gas is completely ideal.

ideal solution A solution in which there is no chemical interaction between the solvent and the solute molecules.

immiscible Substances that do not mix and form more than one phase when brought together.

In Symbol for the element indium.

incandescence The emission of light by a body that is strongly heated. For example, the filament of an electric light bulb and the "limelight" obtained by heating lime.

indicator A substance that indicates, by a change in its color, the degree of acidity or alkalinity of a solution or the presence of a given substance.

indium Element symbol, In; group 3; soft silvery metal; Z 49; A(r) 114.82; density (at 20°C), 7.31 g/cm^3; m.p., 156.6°C; compounds are toxic; named after the indigo line in its spectrum; discovered 1774; forms alloys with low melting points; compounds used in semiconductors and electric motors.

inert A substance that is either very or completely unreactive. Nitrogen and the noble gases are inert.

inert gases *See* group 8 elements.

inhibitor The reverse of a catalyst, a compound that slows down the rate of a chemical reaction. *See* antiknock for the action of lead(IV) tetraethyl in slowing down the ignition of a petrol-air mixture.

inorganic chemistry The study of the chemistry of all the elements excluding the organic compounds made by carbon (it includes the study of carbon-containing compounds such as carbonates, hydrogencarbonates and carbon dioxide).

insoluble A substance that does not dissolve in a particular solvent under certain conditions of temperature and pressure.

insulator A material that does not conduct energy such as electricity, heat, and sound.

iodides Salts of hydriodic acid.

iodine Element symbol, I; halogen, group 7; black solid, producing iodine vapor (violet); Z 53; A(r) 126.9; density (at 20°C), 4.94 g/cm^3; m.p., 113.5°C; name derived from the Greek *iodes*, "violet"; discovered 1811; essential element in diet; used in disinfectants, photography, halogen light bulbs.

iodoform test Test for the presence of an ethanoyl group in ketones or aldehydes. Pour 2 ml of a solution of iodine in potassium iodide solution into a test tube and carefully add sodium hydroxide until the brown color of the solution has almost gone. Add two drops of the solution to be tested. If a fine yellow precipitate is produced, it confirms the presence of an ethanoyl group in the sample.

ion An electrically charged atom or group of atoms. An atom (or group of atoms) tends to lose or gain one or more electrons to form an ion that has a noble gas configuration.

ion bridge *See* half-cell.

ion-electron equation representation of a chemical reaction showing the gain or loss of electrons from an element or ion as the charge on the element or ion changes during the reaction.

ion-exchange A process in which ions of the same charge are exchanged between ions in a solution and ions in a solid in contact with the solution. *See* Permutit.

ionic bond *See* bond.

ionic compound Compounds consisting of ions held together by strong ionic bonds. They tend to be hard solids. Their melting points are high because of the strength of the ionic bonds. Ionic compounds are electrolytes; their ions can move when the compound is melted or dissolved in a suitable solvent.

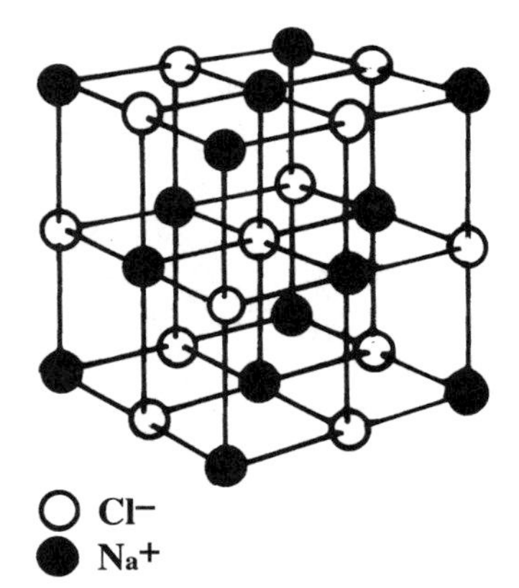

Cl^-
Na^+

Ionic crystal

ionic crystal A type of crystal where ions of two of more elements form a regular three-dimensional arrangement (crystal structure). The ions are held strongly in place by ionic bonds between positive and negative ions.

$OH^-(aq)+H^+(aq) \rightarrow H_2O(l)$

Ionic equation

ionic equation A representation of a chemical reaction where the reactants and products taking part in a reaction are shown as ions. Other ions may be present but do not take part; they are known as spectator ions. For example, in the reaction between nitric acid and sodium hydroxide, forming sodium nitrate and water:

$$HNO_{3(aq)} + NaOH_{(aq)} = NaNO_{3(aq)} + H_2O_{(l)}$$

If this is written in ionic form

$$H^+ NO_3^-{}_{(aq)} + Na^+ OH^-{}_{(aq)} = Na^+NO_3^-{}_{(aq)} + H_2O_{(l)}$$

the sodium and the nitrate ions appear on both sides of the equation, and are said to be spectator ions and the ionic equation is

$$H^+{}_{(aq)} + OH^-{}_{(aq)} = H_2O_{(l)}$$

ionic formula This shows the charges of the ions in an ionic substance, for example, Na^+Cl^- or $Ca^{2+}(Cl^-)_2$.

ionic lattice An ionic crystal of two or more elements that is held together by the electric forces (ionic bonds) between negative and positive ions in a regular structure.

ionic theory Substances that separate into oppositely charged particles (ions) when electricity is passed through a solution of the substance, or a molten sample of the substance.

ionization The process by which an atom becomes an ion by either losing or gaining one or more electrons.

ionization energy The energy needed to remove completely an electron from a neutral gaseous atom or ion against the attraction of the nucleus. The energy needed to remove the first electron is the first ionization energy; the energy for the second is the second ionization energy. These become progressively larger.

ion migration The movement of ions in an electrolyte under the influence of an applied voltage. If ions forming compounds of certain colors are used, it is possible to study the process of ion migration by observing color changes in different parts of the electrolyte (or in filter paper soaked in an electrolyte).

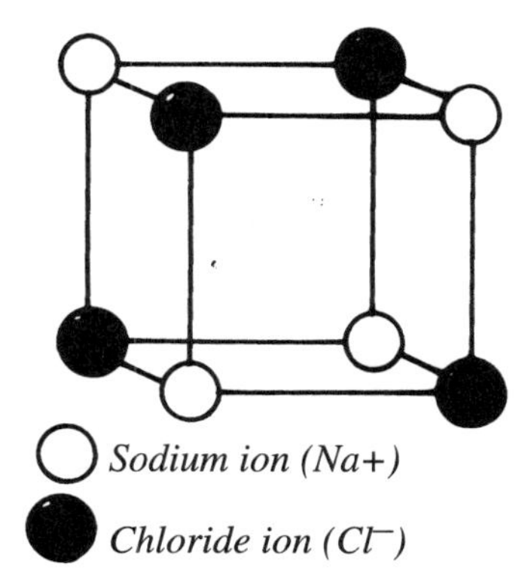

Ionic lattice

Ir Symbol for the element iridium.

iridium Element symbol, Ir; transition element; silvery metal; Z 77; A(r) 192.22; density (at 20°C), 22.42 g/cm^3; m.p., 2410°C; name derived from the Latin and Greek word *iris*, "rainbow"; discovered 1803.

iron Element symbol, Fe; transition element; silvery malleable and ductile metal; Z 26; A(r) 55.85; density (at 20°C), 7.87 g/cm^3; m.p., 1535°C; Old English name *iren*; Latin name *ferrum*; known since prehistoric times; used in construction, usually converted to steel for strength.

iron chloride $FeCl_2$ (iron(II) chloride, ferrous chloride) Anhydrous iron(II) chloride is a white solid. It is deliquescent and becomes green-yellow on absorbing water; m.p., 670°C; r.d., 3.16. It also exists as $FeCl_2.2H_2O$, a green crystalline compound; r.d., 2.36; and as $FeCl_2.4H_2O$, a blue crystalline compound; r.d., 1.93.

$FeCl_3$ (iron(III) chloride, ferric chloride) Anhydrous iron(III) chloride is a black-brown solid; m.p., 306°C; r.d., 2.9. It forms a brownish yellow solution. It also exists as $FeCl_3.6H_2O$, a brown-yellow deliquescent crystalline substance.

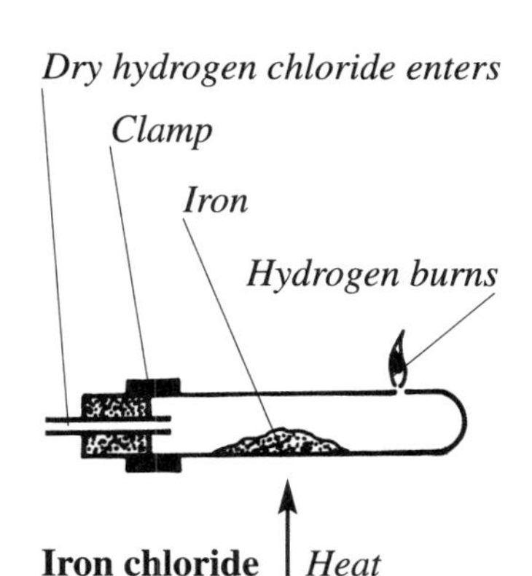

Iron chloride

iron(II) disulfide FeS_2. A yellow crystalline mineral occurring in many rocks. It is often mistaken for gold.

iron—extraction of *See* blast furnace.

iron hydroxide $Fe(OH)_2$ (iron(II) hydroxide, ferrous hydroxide) A solid that is white when pure. It is formed as a dirty green precipitate when an alkali is added to an aqueous solution containing Fe^{2+} ions.

$Fe(OH)_3$ (iron (III) hydroxide, ferric hydroxide) A rust-colored gelatinous solid precipitated by adding an alkali to a solution containing Fe^{3+} ions.

iron oxides FeO (iron(II) oxide, ferrous oxide) r.d., 5.7; m.p., 369°C. A black solid formed by heating iron(II) ethanedioate (FeC_2O_4) in a vacuum. It dissolves in dilute acids.

Fe_2O_3 (iron(III) oxide, ferric oxide) A red-brown insoluble solid, r.d., 5.24, m.p., 1565°C. It is formed when $Fe(OH)_3$ is heated. For the hydrated form ($Fe_20_3.\times H_20$), *see* rust.

(Fe_3O_4 iron(II) di-iron(III) oxide, tri iron tetroxide) A magnetic oxide of iron; it occurs naturally as magnetite. It is formed as a black solid when steam is passed over red-hot iron.

iron pyrites *See* iron(II) disulfide.

iron sulfates $FeSO_4.H_2O$ (iron(II) sulfate, ferrous sulfate) An off-white crystalline compound; r.d., 2.97.

$FeSO_4.7H_2O$. A blue-green crystalline compound; r.d., 1.898 m.p., 64°C; known as green vitriol or copperas.

$Fe_2(SO_4)_3$ (iron(III) sulfate, ferrous sulfate) A yellow hygroscopic crystalline compound; r.d., 3.097. It decomposes above 480°C.

isocyanates Substances that contain the group –N=C=O.

isomers Different (usually organic) compounds having the same molecular formula and relative molecular mass but some different properties, as they have different three-dimensional structures. *See* structural isomer, stereoismer.

isometric **1.** In the study of crystals, *isometric* describes cubic crystal systems where the axes are perpendicular to each other.

2. In the graphic representation of temperature pressure and volume, an isometric line shows how the temperature and pressure of a gas relate to each other at constant volume.

isomorphism The existence of two or more different substances (isomorphs) that have the same crystal structure.

isotonic Solutions that have the same osmotic pressure.

isotope Atoms of the same element (all chemically identical) having the same atomic number but containing different numbers of neutrons, giving different mass number. Some elements occur naturally as a mixture of different isotopes. All elements can produce radio isotopes artificially.

isotropic A substance whose physical properties do not change with direction, such as a cubic crystal.

K Symbol for the element potassium.

kaolin *See* china clay.

Kelvin scale A temperature scale that has no negative values. Its lower fixed point is absolute zero. The size of the unit, the kelvin, is the same as the degree Celsius, and the triple point of water is 273.16 K.

kerosene (kerosene, paraffin oil) A mixture of hydrocarbons mainly consisting of alkanes with between 10 and 16 carbon atoms. It is used in jet engine fuel and in paraffin heaters. It is obtained from petroleum refining. It boils between 160 and 250°C.

keto- A prefix denoting that the substance contains a carbon atom attached to an oxygen atom by a double bond and to two other carbon atoms by single bonds, i.e., it contains a carbonyl group.

ketone A family of organic compounds that contain two organic radicals connected to a carbonyl group. Names have the suffix -one. General formula: R–CO–R, where R represents an aliphatic or aromatic hydrocarbon group. Ketones are very reactive.

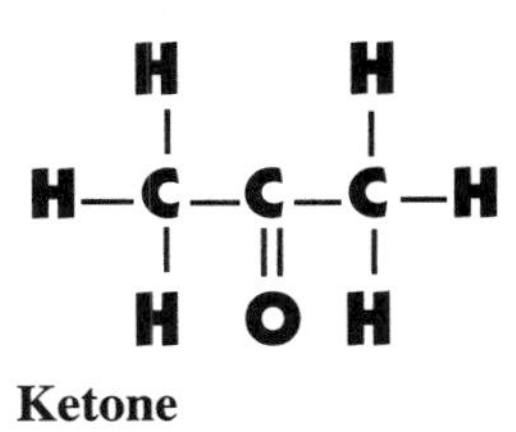

Ketone

kinetic theory All matter consists of particles, such as atoms, ions, or molecules, which are in a state of continual motion, and this motion is dependent on temperature. The theory explains some of the physical properties of materials, and it is particularly useful in explaining the behavior of gases, because gas molecules are relatively unaffected by forces of attraction between molecules. In solids and liquids, the amount of movement is less, due to bonds between particles.

Kr Symbol for the element krypton.

krypton Element symbol, Kr; noble gas, group 8; Z 36; A(r) 83.3; density (at 20°C), 3.743 g/l at STP; m.p., –157.3°C; name derived from the Greek *kryptos*, "hidden"; discovered 1898; used in fluorescent lights.

La Symbol for the element lanthanum.

lactose A disaccharide with the formula $C_{12}H_{22}O_{11}$; it is an isomer of maltose. Lactose is a white crystalline solid occurring in the milk of mammals. It is a reducing carbohydrate.

lanthanides (lanthanoids, lanthanons, rare earths) A series of elements comprising lanthanum, cerium, praesodymium, neodymium, promethium, samarium, europium, gadolinium, terbium, dysprosium, holmium, erbium, thulium, ytterbium, and lutetium. As their outer electronic structure is very similar (the f orbital in their fourth shell is being filled), they have similar chemical properties. The metals are shiny and are attacked by water and acids. Their usual oxidation number is +3.

lanthanum Element symbol, La; lanthanide; silvery metal; Z 57; A(r) 138.91; density (at 20°C), 6.15 g/cm^3; m.p., 921°C; name derived from the Greek *lanthanein*, "to lie unseen"; discovered 1839; used in alloys with magnesium and aluminum and in steels.

latent heat The amount of heat that is absorbed or released by a substance during a change of state (fusion or vaporization) at constant temperature.

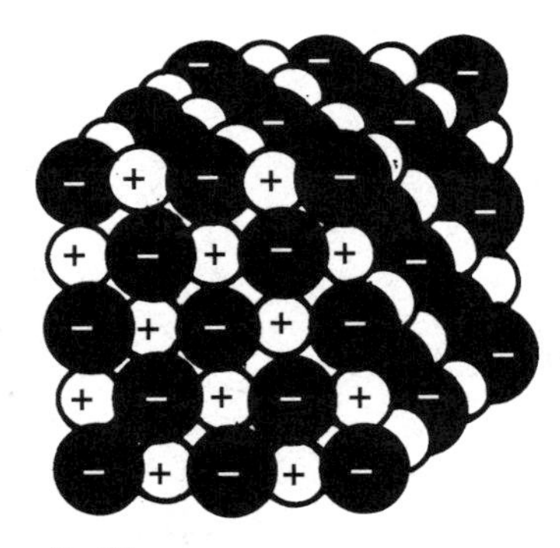
Lattice

lattice The orderly three-dimensional arrangements of atoms, molecules, or ions seen in crystals.

law A rule describing certain natural observable phenomena or the relationship between effects of variable quantities.

law of combining masses Elements combine in the ratio of their combining masses or in a simple multiple of that ratio.

law of constant composition or **law of definite proportions** or **Proust's law** A pure chemical compound always contains the same elements combined in the same proportions by mass.

law of definite proportions *See* law of constant composition.

law of mass action This applies to a reversible chemical reaction such as aA + bB $\rightleftharpoons$ cC+dD at chemical equilibrium where the rate of the forward reaction is equal to the rate of the back reaction.
The equilibrium constant

$$K = \frac{[C]^c[D]^d}{[A]^a[B]^b}$$

where [A] etc. are the active masses of the substances. These are often taken as their molecular (molar) concentrations. For gas reactions, partial pressures are used rather than concentrations.

The equilibrium constant shows the position of equilibrium; if it has a low value, it shows that [C] and [D] are low compared with [A] and [B]. It also indicates how the equilibrium would shift if one of the concentrations changed.

law of multiple proportions If two elements (A and B) combine to form more than one compound, the different masses of A that combine with a fixed mass of B are in a simple ratio.

H	Li	Be	B	C	N	O
F	Na	Mg	Al	Si	P	S
Cl	K	Ca	Cr	Ti	Mn	Fe
(doh	ray	me	fah	soh	lah	te)

Law of octaves

law of octaves or **Newlands' law** An arrangement of 56 elements in order of ascending atomic weight made by Newlands in 1863. He found that if the elements were arranged in order of ascending atomic weight and placed in a table so that the first eight were aligned above the second eight (and so on), the elements in a column had similar properties.

law of reciprocal proportions *See* law of combining masses.

lawrencium Element symbol, Lr; actinide; Z 103; A(r) 256; no solid compounds are known; named in honor of the physicist Ernest O. Lawrence; discovered 1961.

laws of chemical combination The laws representing the way in which elements combine in the formation of chemical compounds. There are three laws: the law of constant composition, the law of multiple proportions, and the law of combining masses.

lead Element symbol, Pb; group 4; silvery white metal; Z 82; A(r) 207.2; density (at 20°C), 11.35 g/cm^3; m.p., 327.5°C; compounds are toxic; Old English name *lead*; Latin name *plumbum*; known since prehistoric times; used in batteries and in water, noise, and radiation shielding. It is also used in high-quality glass production and the manufacture of storage batteries.

lead bromide $PbBr_2$. A white crystalline poisonous solid; m.p., 373°C. It is almost insoluble in cold water, but fairly soluble in hot water.

lead carbonate $PbCO_3$. A poisonous white salt that is insoluble in water. It occurs naturally in the mineral cerussite and is used as a pigment.

lead(II) carbonate hydroxide $2PbCO_3.Pb(OH)_2$ (white lead, lead(II) carbonate hydroxide) It occurs naturally in hydroxycerussite. It decomposes at 400°C. White lead was used widely in paint, but it discolored on contact with hydrogen sulfide in the atmosphere. Fears of lead poisoning also led to the decline in use of white lead in paint.

lead chamber process *See* chamber process.

lead chloride $PbCl_2$. A white crystalline solid that is almost insoluble in cold water. It is fairly soluble in hot water.

lead iodide PbI_2. It is formed as a golden yellow precipitate in solution.

lead nitrate $Pb(NO_3)_2$. Lead nitrate exists as colorless crystals that are used as a mordant in dyeing and in the manufacture of chrome-yellow pigment.

lead oxides PbO (lead(II) oxide, lead monoxide) A yellow compound that is amphoteric and insoluble in water; m.p., 886°C. It exists in two crystalline forms: litharge, obtained when the oxide is heated above the melting point; and massicot, when it is heated to temperatures below the melting point. It is used in the manufacture of paint and glass.

PbO_2 (lead(IV) oxide, lead dioxide) A dark brown solid that decomposes to form lead(II) oxide and oxygen on heating. It is a powerful oxidizing agent. It is used in lead-acid cells.

$Pb_3O_4.(Pb_2PbO_4)$ (di-lead(II) lead(IV) oxide, tri-lead tetroxide, red lead) A red solid produced on heating lead(II) oxide in oxygen at 400°C. It decomposes at 500°C to form lead(II) oxide and oxygen. It is used in paint manufacture and in the manufacture of lead(IV) oxide.

lead sulfate $PbSO_4$. A white insoluble crystalline solid; m.p., 1170°C. It has been used in paint because it is less susceptible to discoloration than lead(II) carbonate hydroxide, but it is now less used because of fears about lead poisoning.

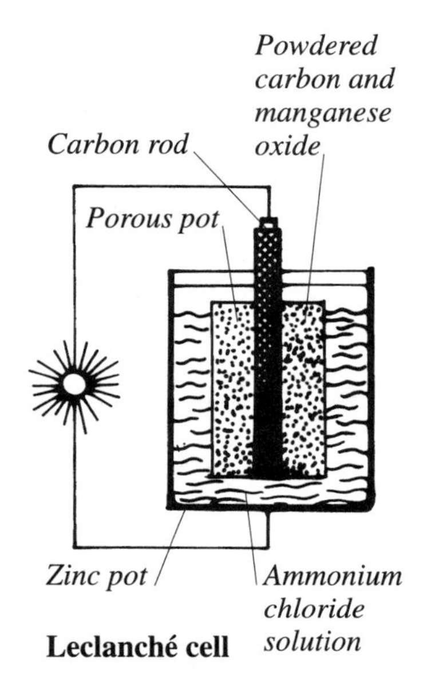

Leclanché cell

lead sulfide PbS. A brownish black insoluble crystal; m.p., 1114°C; occurs naturally as the mineral galena.

Le Chatelier's principle If a chemical reaction is at equilibrium and a change is made to any of the conditions, further reaction will take place to counteract the changes in order to re-establish equilibrium.

Leclanché cell A primary cell with an e.m.f. of 1.5 volts and internal resistance of 1 ohm. The positive electrode is a carbon rod that is surrounded by a mixture of powdered carbon and manganese dioxide in a porous pot. The pot stands in ammonium chloride solution contained in a zinc pot, which forms the negative electrode.

Li Symbol for the element lithium.

ligand An atom (or group of atoms) surrounding the central atom in a complex.

lime *See* calcium oxide.

limestone *See* calcium carbonate.

limewater A solution of calcium hydroxide that is used to test for the presence of carbon dioxide. If carbon dioxide is bubbled through limewater, a solid precipitate of calcium carbonate is formed.

linear molecules Molecules whose atoms are in a line. For example, ethyne, carbon dioxide.

lipid General name for a loosely defined group of organic compounds that are considered to be the esters of long-chain carboxylic acids and various alcohols. Oils, fats, and waxes are lipids.

liquid A state of matter between solid and gas. Particles are loosely bonded, so can move relatively freely. A liquid has low compressibility.

lithium Element symbol, Li; alkali metal, group 1; silvery white metal; Z 3; A(r) 6.94; density (at 20°C), 0.53 g/cm^3; m.p., 180.5°C; reacts with water; name derived from the Greek *lithos*, "stone"; discovered 1817; used in low-melting alloys; compounds have many uses, including ceramics and fungicides.

lithium carbonate Li_2CO_3. A white solid; m.p., 735°C. It decomposes above 1200°C. Lithium carbonate is used in the prevention and treatment of manic depressive disorders.

lithium chloride LiCl. A very deliquescent soluble white solid that is used in mineral waters and as a flux in soldering.

lithium oxide Li_2O (lithia). A white crystalline compound; m.p., 1700°C; used in lubricating greases, ceramics, and glass.

litmus A soluble purple compound extracted from lichens. It can be used as an acid-base indicator because its color changes are dependent on the pH of a solution with which it is mixed. It is red in acid solutions (pH less than 5) and blue in alkaline solutions (pH more than 8). Paper that has been soaked in a solution of litmus and dried is called litmus paper.

lone pair A pair of electrons in a filled atomic orbital in the outermost shell of an atom. They are not involved in the formation of covalent bonds but have an effect on the shape of molecules (by affecting the angles of the bonds formed) and also cause the molecule to become strongly polar. *See* polar molecule.

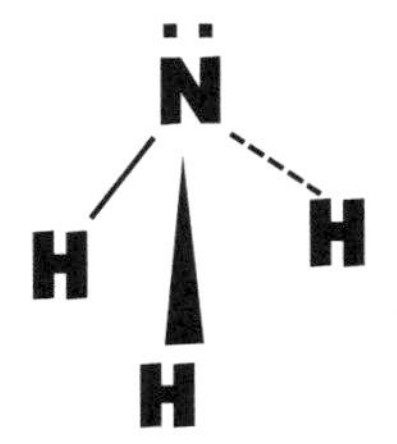

Lone pair of electrons in ammonia

Lr Symbol for the element lawrencium.

Lu Symbol for the element lutetium.

luminescence Light emission from a substance caused by an effect other than heat. Fluorescence and phosphorescence are forms of luminescence.

lutetium Element symbol, Lu; rare earth/lanthanide; silvery metal; Z 71; A(r) 174.97; density (at 20°C), 9.84 g/cm^3; m.p., 1663°C; named for *Lutetia*, the Latin name for Paris; discovered 1907; used as a catalyst.

M *See* molarity.

macromolecules Very large molecules, such as polymers and proteins, whose relative molecular mass is larger than 1000.

magnesite Mineral containing magnesium carbonate.

magnesium Element symbol, Mg; alkali earth metal, group 2; silver-white metal; Z 12; A(r) 24.31; density (at 20°C), 1.74 g/cm^3; m.p., 648.8°C; will burn in air; named after Magnesia, a district in Thessaly; discovered 1808; used in alloys and castings. Some compounds have medicinal uses; also used in sugar, cement, paper manufacture and many other industries.

magnesium carbonate $MgCO_3$. A white compound that is slightly soluble in water. It is used in making magnesium oxide and as a drying agent (particularly in table salt as an anti-caking agent). It is used medically as laxative and antacid.

magnesium chloride $MgCl_2$. A white solid compound; m.p., 714°C; b.p., 1412°C. It is deliquescent, forming the hexahydrate $MgCl_2.6H_2O$. Magnesium chloride is used in fireproofing and fire-extinguishing materials, and in the textile industry.

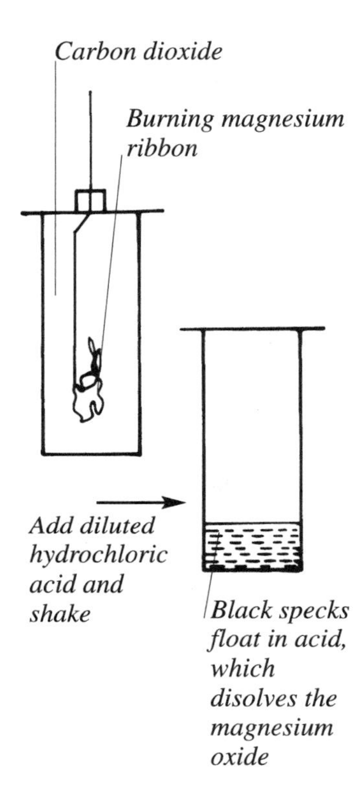

Magnesium oxide

magnesium hydrogen carbonate $Mg(HCO_3)_2$. A soluble magnesium salt that is only stable in solution. It decomposes on heating to form magnesium carbonate, carbon dioxide, and water. It is formed by the action of carbon dioxide and water on calcium carbonate and is one of the causes of temporary hardness in water (*see* hardness in water).

magnesium hydroxide $Mg(OH)_2$. A slightly soluble crystalline white powder that decomposes at 350°C. It occurs naturally in the mineral brucite or it can be prepared by adding a strong alkali to a solution containing magnesium ions. It is used medically as an antacid (milk of magnesia).

magnesium oxide MgO (magnesia). A white solid (m.p., 2800°C) occurring naturally in the mineral periclase. It is prepared industrially by the thermal decomposition of magnesite. It has many uses, including reflective coatings and as a component of semiconductors.

magnesium sulfate $MgSO_4$. A white soluble salt that exists in both anhydrous form and in hydrated crystalline form. $MgSO_4.H_2O$ magnesium sulfate monohydrate is found as the mineral kieserite. The heptahydrate ($MgSO_4.7H_2O$) is found as the mineral epsomite and is known as Epsom salts. Epsom salts are used medically as a laxative, in the manufacture of fertilizers and matches, in sizing and fireproofing textiles, and in tanning leather.

magnetite Fe_3O_4. A mineral containing iron(II) di-iron(III) oxide. It is a black solid that is a natural magnet or "lodestone."

malachite A copper ore formed of hydrated copper carbonate. Its formula can be written either $Cu_2(OH)_2CO_3$ or $CuCO_3.Cu(OH)_2$. Its color ranges from dark to light green, and the mineral is found with these colors forming striking patterns of bands.

malleability of metals Metals can be rolled into flat sheets and hammered into different shapes because they are malleable.

maltose A disaccharide with the formula $C_{12}H_{22}O_{11}$. An isomer of lactose, it is a reducing carbohydrate.

manganese Element symbol, Mn; transition element; soft gray metal; Z 25; A(r) 54.94; density (at 20°C), 7.3 g/cm^3; m.p., 1244°C; name derived from medieval translation of the Latin *magnesia*; discovered 1774; used in steel manufacture and in alloys. Compounds have many industrial uses.

manganese (IV) oxide MnO_2 (manganese dioxide). It is an insoluble black powder that is made by heating manganese(II) nitrate. It is a powerful oxidizing agent.

margarine *See* hydrogenation.

marsh gas More commonly known as methane (CH_4), it is found in marshy districts, where it is formed by the anaerobic bacterial decomposition of vegetable (and animal) material.

mass number The total number of protons and neutrons (nucleons) in the nucleus of an atom.

Md Symbol for the element mendelevium.

meitnerium Element symbol, Mt; transition element; Z 109; A(r) 266; named in honor of Austrian physicist Lise Meitner; discovered 1982. Formerly known as unnilennium.

melamine $C_3H_6N_6$ (cyanuramide, 2,4,6-triamino-1,3,5-triazine). A white or colorless crystalline compound (m.p., 354°C) produced from urea. It forms strong thermosetting polymers, which are stable to heat and light, on undergoing condensation polymerization with methanal.

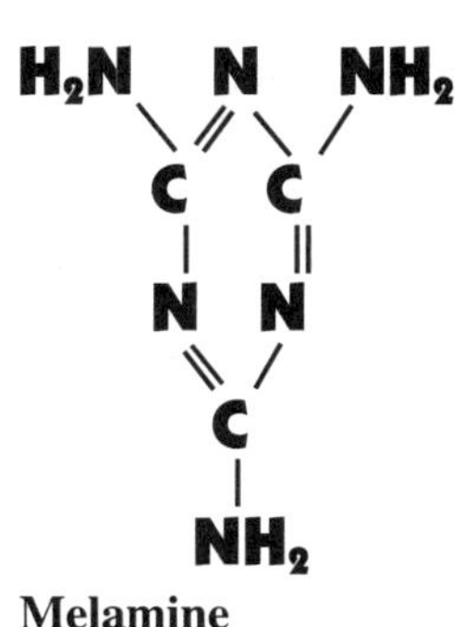

Melamine

melting The change of state from solid to liquid. It occurs when the particles in the solid lattice have gained sufficient energy to break the bonds that hold them in the lattice.

Lattice (solid)

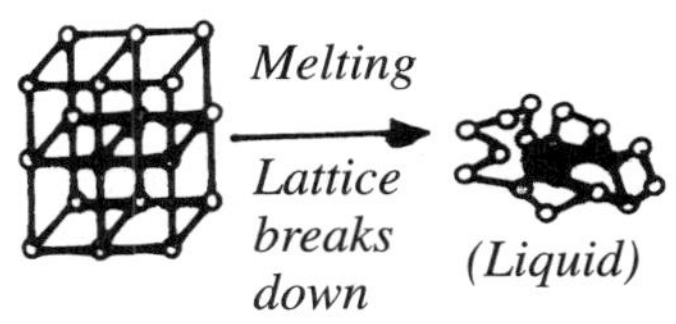

Melting

Mendeleyev classification Scientists in the 19th century found that there were elements that had similar properties, and formulated the idea that there were families or groups of elements. The Russian chemist Mendeleyev suggested in 1869 that "The properties of the elements are in periodic dependence upon their atomic weights" and related the chemical properties of each group to those of the other groups. In his table of elements, elements in the same group were placed in the same vertical column and these columns were arranged in the order given by the elements' gradual change in chemical reactivity and increasing atomic weight.

mendelevium Element symbol, Md; actinide; Z 101; A(r) 258; no solid compounds known; named in honor of Dmitri Mendeleyev, Russian chemist; discovered 1955.

mercury Element symbol, Hg; transition element; silver liquid metal; Z 80; A(r) 200.59; density (at 20°C), 13.55 g/cm^3; m.p., –38.8°C; mercury and many of its compounds are toxic and tend to accumulate in the bodies of higher animals; named for the planet Mercury, Latin *hydrargyrum*, "liquid silver"; known since prehistoric times; used in dental fillings, thermometers.

mercury cathode cell *See* Castner-Kellner cell.

mercury(II) oxide HgO (mercuric oxide). A red solid when formed by heating mercury in oxide, a yellow precipitate formed by the addition of a strong alkali to a solution containing mercury(II) ions. The color difference is caused by particle size. Both forms decompose to form mercury and oxygen when strongly heated.

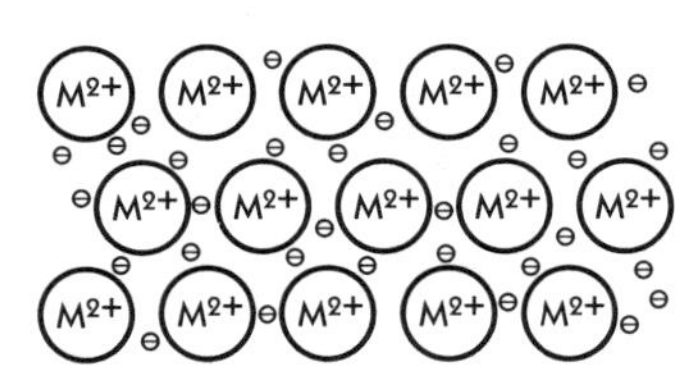

Metallic bond

metallic bond The bonding formed in metallic crystals where there is a lattice formation of positively charged ions within a sea of electrons that binds them together. There are insufficient electrons in this "sea" to form individual bonds between atoms. Atoms that have more electrons in their outer shell can contribute more electrons to the electron sea. The bonding is thus stronger and the metallic crystal formed is harder and denser as the ions are held more tightly. Thus alkali metals are softer and less dense than transition elements.

metallic crystal Metals form a giant crystalline structure with high melting and boiling points. The structure is formed of metal ions. Metal atoms have excess electrons in their outer shell (sodium has one, calcium two); these are mobile within the giant structure of ions and allow the solid to be a good conductor.

metalloid An element that has both metallic and nonmetallic properties.

metal properties Metals are strong, dense, and malleable, and they are good conductors of heat and electricity.

metals and acids All metals higher in the reactivity series than copper react with a dilute acid to form a salt and hydrogen.

metals and oxygen All metals higher in the reactivity series than silver combine with oxygen when heated.

metals and water All metals higher in the reactivity series than aluminum react with water to give hydrogen gas and the metal hydroxide.

metals—extraction of Most metals are not found naturally as the pure metal, exceptions being very unreactive metals such as gold and silver. Most metals are found combined with other elements in ores. Ores have to be processed to extract the metal. An example of this is the extraction of iron from iron ore (where iron is present as Fe_2O_3) in the blast furnace (*see* blast furnace). Metal oxides are reduced to the metal during extraction.

metals recycling Waste steel and aluminum can be recycled. This reduces energy use because making an object from recycled aluminum uses about 5% of the energy that would be used if new aluminum were used. Recycling is also of value because of the finite nature of the world's resources of metals.

alu

Metal recycling

metastable Describes a system that appears to be stable but that can undergo a rapid change if disturbed. For example, if water is slowly cooled below 0°C (supercooled water), it appears to be a stable liquid, but if a piece of ice is added, the water freezes rapidly as the system attains a lower energy state. This condition is also seen in supersaturated solutions.

methanal HCHO (formerly formaldehyde) The simplest aldehyde. It is a colorless gas with a pungent smell; m.p., –92°C; b.p., –21°C.

methane CH_4. The simplest alkane. A colorless, tasteless, odorless flammable gas; m.p., –182°C; b.p., –162°C. It is found in natural gas and in coal gas and can be formed by the anaerobic decomposition of vegetable and animal compounds. It is used as a fuel. It is slightly soluble in water, forming a neutral solution. Methane is used in the formation of many organic compounds.

methanoic acid HCOOH (formic acid) The simplest carboxylic acid. A colorless liquid with a pungent smell. It is used in textile dyeing, electroplating and in some pesticides. Salts of methanoic acid are called methanoates (formerly formates).

methanol CH_3OH (methyl alcohol, wood alcohol) The simplest alkanol. It is a volatile, colorless, flammable, poisonous liquid that is produced by reacting methane with steam over a nickel catalyst at high temperature and pressure. This produces synthesis gas, which is converted to methanol when passed over a zinc oxide and chromium(III) oxide catalyst at 300°C under pressure. Methanol is used as a solvent and in methanal production.

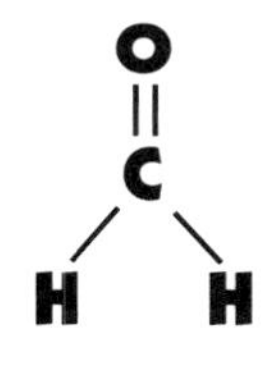

Methanal

methyl alcohol *See* methanol.

methylated spirits A form of ethanol (90% of the mixture) that has been rendered unfit for human consumption (and thus escapes taxation) by the addition of methanol (9.5%) and pyridine (0.5%) and a trace of blue dye. It is used as a solvent.

methylbenzene $C_6H_5CH_3$ (toluene) An aromatic hydrocarbon found in coal tar. It is a colorless insoluble flammable liquid used as a solvent and in the synthesis of other organic compounds.

methyl group or **radical** The organic group CH_3^-.

methyl orange Water-soluble acid-base indicator. It is red in solutions where the pH is less than 3.2 and yellow in solutions having a pH above 4.4. Between 3.2 and 4.4 it is orange. It is used in titrations of a weak base with a strong acid, giving a weak acidic end point.

methyl tertiary butyl ether (MTBE) An additive in unleaded gasoline.

Mg Symbol for the element magnesium.

mineral A natural inorganic substance with distinct chemical composition and internal structure. Various kinds of minerals form the ingredients of rocks. Quartz is the most plentiful rock-forming mineral.

mineral processing The processes by which elements found in the Earth's crust in minerals and metallic ores are changed into more useful forms.

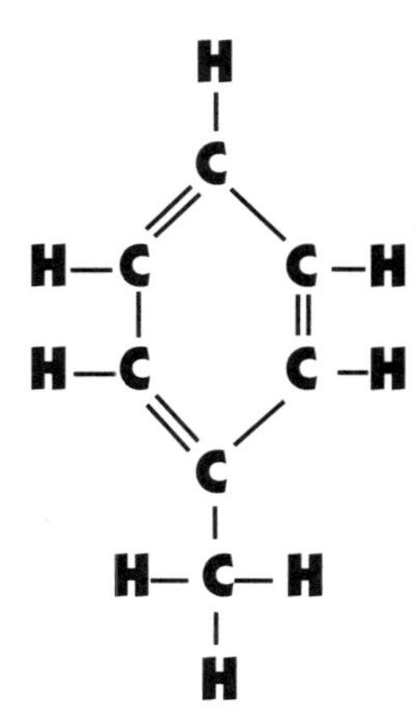

Methylbenzene

The mineral ore is extracted, crushed, and then purified by chemical means. For examples of purification, *see* blast furnace (for the smelting of iron ore); Castner-Kellner process (for the production of chlorine); Frasch process (for the production of sulfur).

miscible Liquids that mix together completely. The result of mixing looks like a single liquid and requires fractional distillation to separate the liquids.

mixture A system that consists of two or more substances (solid, liquid, or gas) present in any proportions in a container. There is no chemical bonding between substances. A mixture can be separated using physical methods. The formation of a mixture does not involve a change in temperature.

Mn Symbol for the element manganese.

Mo Symbol for the element molybdenum.

Mohs' scale A scale that measures the hardness of minerals by their ability to scratch one another. A mineral is given a number on Mohs' scale according to its ability to scratch one of the reference materials. In order of increasing hardness: 1, talc; 2, gypsum; 3, calcite; 4, fluorite; 5, apatite; 6, feldspar; 7, quartz; 8, topaz; 9, corundum; 10, diamond.

molality (m) Concentration of solution giving the number of moles of solute dissolved in 1 kg of solvent.

molar gas constant or **universal gas constant** (R) It is used in the gas equation $PV=nRT$. Its value is 8314 JK^{-1} $mole^{-1}$.

molarity (M) Concentration of solution giving the number of moles of solute dissolved in 1 dm^3 of solution.

molar solution A solution containing one mole of a solute in one liter of solution.

molar volume (gram molecular volume) (V_m) The volume occupied by one mole of a substance.

molar volume of gas At STP all gases have approximately equal molar volumes, 22.4 cubic decimeters.

mole The amount of a substance that contains the same number of entities (atoms, molecules, ions, any group of particles), but the type must be specified, as there are atoms in 0.012 kg of the carbon-12 isotope. The actual number is known as the Avogadro number. Its value is 6.023×10^{23}.

molecular crystal A type of crystal where molecules form a regular three-dimensional arrangement (crystal structure). The atoms within the molecule are held firmly in place but the molecules are held together by weak bonds, such as van der Waal's bonds and hydrogen bonds, and are easily separated. Molecular crystals have low melting points and they do not conduct electricity (there are no mobile electrons in the structure). Organic compounds tend to form molecular crystals.

molecular formula This indicates both the type of atom present (using the symbols that represent each element in the periodic table) and the number of each atom in the molecule. The molecular formula may be a multiple of its empirical formula.

Molecular formula

molecularity The number of molecules taking part in a chemical reaction, which form an activated complex during one step of a process. A reaction is unimolecular if one molecule takes part, bimolecular if two molecules take part, and so on. *See* order of reaction.

molecular mass or **weight** *See* formula mass.

molecular orbitals The orbitals belonging to a group of atoms forming a molecule. Only the outer electrons are usually considered as forming molecular orbitals.

In the formation of a molecule, the valence electrons (electrons that make the bond between the atoms) are affected by both nuclei, and they move in molecular orbitals whose shape is governed by the shape of the individual atomic orbitals. For example, in a bond between two hydrogen atoms, each of the electrons that is to form the bond between the atoms is in an s orbital; these s orbitals overlap to form a molecular orbital between the two nuclei. This orbital is known as a sigma orbital. If bonding takes place between p orbitals, the bond is in two parts (at the end of each lobe of the p orbital) and is known as a pi orbital.

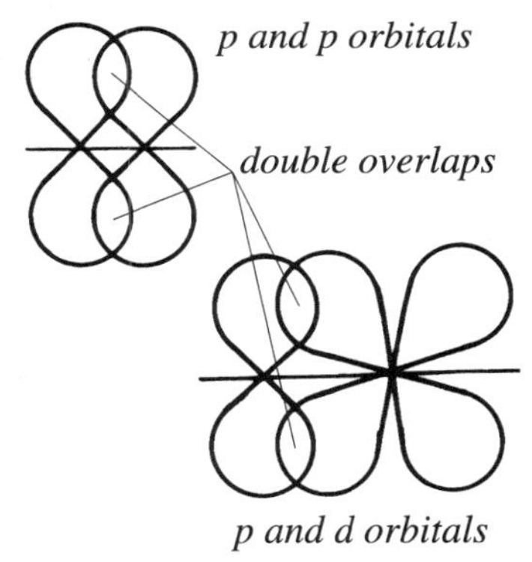

Molecular orbitals

The shape of molecular orbitals can be seen in terms of hybrid orbitals, when the electrons forming a bond are in different types of orbitals in a shell. Carbon, for example, has one s orbital and three p orbitals in its outer shell, and when forming tetrahedral molecules (i.e., making four bonds), it is considered that rather than forming one bond with an electron in an s orbital and three with electrons in p orbitals, it forms four sp^3 hybrid orbitals.

molecule The smallest part of an element or chemical compound that can exist independently with all the properties of the element or compound. It is made up of one or more atoms bonded together in a fixed whole number ratio.

molecules—shapes of The shape of molecules is governed by the arrangement of the bonds within them. For example, the four bonds of carbon are arranged tetrahedrally around the carbon atom and the angle between bonds is about 109.5°. The nitrogen atom in the ammonia molecule forms three covalent bonds with hydrogen atoms, but this leaves an unshared pair of lone electrons around the nitrogen atom. This lone pair repels the shared electrons in the covalent bonds and reduces the angle between the covalent bonds to about 108°. In the water molecule, the oxygen atom has two lone pairs of electrons and these repel the shared electrons in the oxygen-hydrogen bonds even more, reducing the angle between these bonds to 104.5°.

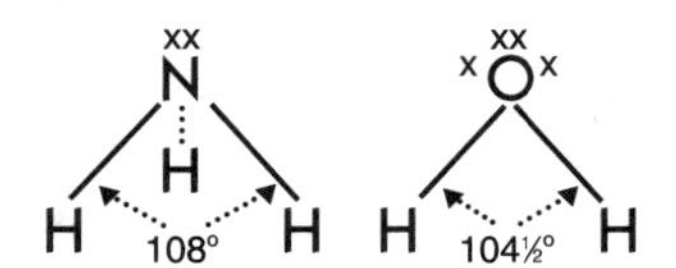

Shapes of molecules

mole fraction A measure of the amount of one of the components in a mixture. It is calculated by dividing the number of moles of the component present in the mixture by the total number of moles of substances present. The sum of the mole fractions of the components of a mixture is one.

molybdenum Element symbol, Mo; transition element; silvery hard metal; Z 42; A(r) 95.94; density (at 20°C), 10.22 g/cm^3; m.p., 2617°C; chemically unreactive; name derived from the Greek *molybdos*, "lead"; discovered 1778; used in steel alloys; compounds are used in pigments.

monatomic molecule A molecule that consists of one atom. Noble gases are monatomic; they have a full outer shell of electrons, and it is difficult for them to share electrons to form bonds with other atoms.

mono- A prefix meaning "one."

monobasic acid An acid that has one replaceable hydrogen atom. Normal salts only can be formed. *See* basicity of acids.

monomer A basic unit from which a polymer is made, either naturally (where glucose [$C_6H_{12}O_6$] is the basic unit of polysaccharides, *see* carbohydrate) or, more usually, synthetically (where a monomer is used in the production of plastic polymers).

monosaccharides *See* carbohydrate.

monotrope Allotropes of an element that exhibit monotropy.

monotropy An element that can exist in more than allotrope but one is always more stable under all conditions. The other forms are metastable (*see* metastable).

monovalent (univalent) Having a valency of one.

mordant A substance used in dyeing to fix the color of the dye onto the fiber.

mortar A mixture of slaked lime and sand made into a paste with water. It sets to a hard mass as the water evaporates and the slaked lime slowly reacts to form calcium carbonate.

Mt Symbol for the element meitnerium.

multiple bonds Covalent bonds that contain more than two electrons. Double bonds contain four electrons. The first two electrons are considered to form a normal bond and the second two electrons are considered to be in p orbitals in the outer shells of the atoms forming the bond. These are brought close enough by the first bond to interact and form pi orbitals (*see* molecular orbitals).

Triple bonds contain six electrons. The first two are considered to form a normal bond and the second two to form a pi orbital (as in the double bond). If the outer shells of the atoms forming the bond each contain another electron (not already forming a bond) in a p orbital, a second pi orbital is formed.

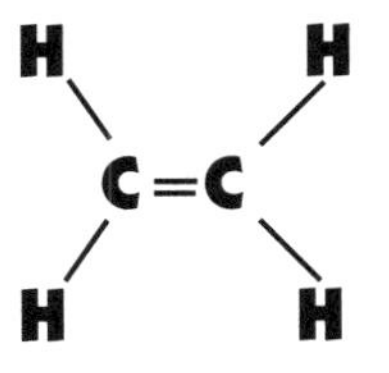

pi bond

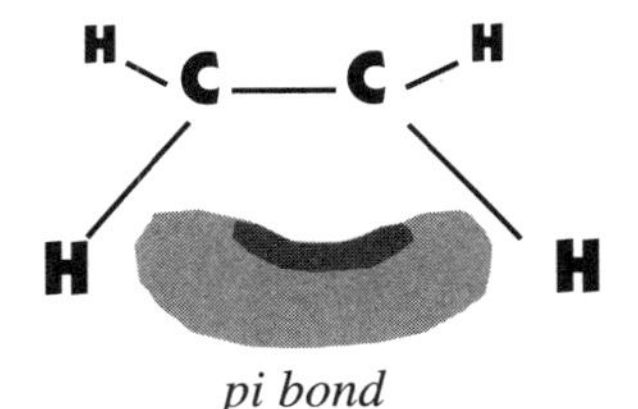

pi bond

Multiple bonds

N Symbol for the element nitrogen.

Na Symbol for the element sodium.

naphtha A mixture of hydrocarbons produced from petroleum by fractional distillation. It is the fraction collected between 80–160°C. It is then converted into smaller molecules by cracking.

naphthalene $C_{10}H_8$. An aromatic organic compound formed of two fused carbon rings. It is a white crystalline solid that sublimes at low temperatures; m.p., 80°C; b.p., 218°C. It is used to produces dyes and plastics.

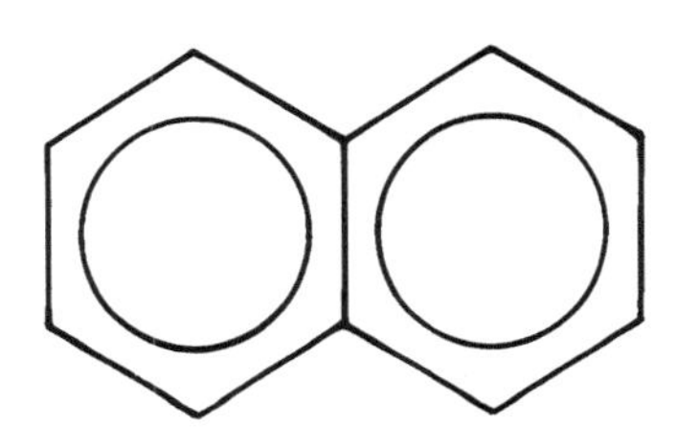

Naphthalene

native An element that is found naturally in its free state, uncombined with other elements.

natural gas *See* fossil fuels.

Nb Symbol for the element niobium.

Nd Symbol for the element neodymium.

Ne Symbol for the element neon.

neodymium Element symbol, Nd; rare earth element/lanthanide; soft silvery metal; Z 60; A(r) 144.24; density (at 20°C), 7 g/cm^3; m.p.,1021°C; name derived from the Greek words *neo*, "new," and *didymos*, "twin"; discovered 1885; used in glass coloring (violet-purple).

neon Element symbol, Ne; noble gas, group 8; colorless gas; Z 10; A(r) 20.18; density (at 20°C), 0.9 g/l at STP; m.p., –248.7°C; forms no normal chemical compounds; name derived from the Greek *neos*, "new"; discovered 1898; used in red fluorescent tubes.

neptunium Element symbol, Np; actinide; radioactive metal; Z 93; A(r) 237.051; density (at 20°C), 20.45 g/cm^3; m.p., 640°C; named for the planet Neptune; discovered 1940.

neutral a solution whose pH is 7. It is neither acidic nor alkaline.

neutralization The reaction of an acid and a base forming a salt and water. The properties of acids and bases disappear when the reaction is complete, at the end point. The solution is neutral.

neutralizers Substances (such as alkalis and carbonates) added to neutralize acid conditions. Lime is added to acid soil, to lower its pH, and to lakes where the water is becoming too acidic, to support plant and animal life. Indigestion tablets also contain neutralizers to neutralize the excessive stomach acid life that is causing indigestion.

neutral oxides Oxides (such as carbon monoxide and dinitrogen oxide) that have neither the properties of an acid nor a base.

neutron One of the three basic particles in an atom, it is found in the nucleus. With the proton, it is one of the most massive of the subatomic particles. It has zero charge.

neutron number The number of neutrons in the nucleus of an atom. All isotopes of an element have the same atomic number but different neutron numbers.

Ni Symbol for the element nickel.

nichrome A group of nickel-chromium alloys that have good resistance to oxidation.

nickel Element symbol, Ni; transition element; silvery white metal, malleable and ductile; Z 28; A(r) 58.71; density (at 20°C), 8.9 g cm^3; m.p., 1453°C; name derived from the German *Kupfernickel*, “demon’s copper”; discovered 1751; used in coins and in steel alloys; used as catalyst in hydrogenation.

niobium Element symbol, Nb; transition element; soft gray-blue metal, ductile; Z 41; A(r) 92.91; density (at 20°C), 8.6 g/cm^3; m.p., 2468°C; named after Niobe, daughter of Tantalus (Greek mythology); discovered 1801; used in special steels.

nitrates Salts of nitric acid. All metallic nitrates are soluble in water. Nitrates, such as sodium nitrate and ammonium nitrate, are important as fertilizers, although overuse can lead to pollution of water (*see* eutrophic).

nitrates—in fertilizers Plant growth requires nitrogen. Different plants require it in different amounts (cereals require more than potatoes), and nitrogen is removed from soil by growing plants. To ensure growth of plants each year, nitrogen should be added. This can be in the form of a synthetic fertilizer that contains nitrogen in the form of nitrates or soluble ionic compounds that dissolve into the soil where the roots of the crop can use them.

niter *See* potassium nitrate.

nitric acid HNO_3. A colorless, corrosive, poisonous, fuming liquid; r.d., 1.5, m.p., –42°C; b.p., 83°C. It is a strong acid that forms nitrates (soluble salts). Nitric acid is a strong oxidizing agent. It is produced industrially by the Ostwald process. It is used in the manufacture of fertilizers and explosives.

nitrites Salts of nitrous acid. Both sodium and potassium nitrite are formed by heating the corresponding nitrate. They are used in the curing of meat.

nitro-chalk Ammonium nitrate to which powdered chalk has been added to prevent the formation of lumps. It is a fertilizer.

nitrogen Element symbol, N; group 5; colorless gas; Z 7; A(r) 14.01; density (at 20°C), 1.25 g/l at STP; m.p., –209.9°C; name derived from the words *niter* ("saltpeter," 18th century) and *genes* ("producing," Greek); discovered 1772; used in Haber process to synthesize ammonia.

nitrogen dioxide NO_2. A poisonous brown gas prepared from the reaction between concentrated nitric acid and copper or by heating dry lead(II) nitrate. It forms a mixture of nitric and nitrous acids in water.

nitrogen fixing The conversion of atmospheric nitrogen into nitrogenous substances. This occurs naturally by the action of certain soil bacteria on the nitrogen in ammonia, and the conversion of atmospheric nitrogen to its oxides by lightning. The Haber process is an industrial process for the fixing of nitrogen.

nitrogen monoxide NO. A colorless, poisonous gas; m.p., –163.6°C; b.p., –151.8°C. It forms nitrogen dioxide on contact with the oxygen in air.

nitrogen oxides in car exhaust gases In a gasoline engine, the fuel/air mixture is compressed and ignited with a spark. The nitrogen in the air reacts to form nitrogen dioxide, which is emitted and adds to atmospheric pollution because it dissolves in rainwater to increase the problem of acid rain (*see* acid rain).

nitrous acid HNO_2. A weak acid existing only in aqueous solution. It forms salts called nitrites. Nitrous acid decomposes on heating to form nitrogen dioxide and nitric acid.

No Symbol for the element nobelium.

nobelium Element symbol, No; actinide; radioactive metal (most stable isotope 254 has half-life of 55 seconds) Z 102; A(r) 255; named in honor of Alfred Nobel, Swedish inventor and industrialist; discovered 1958.

noble gases *See* group 8 elements.

noble gas structure An atom that has a stable electronic structure. Noble gases have eight electrons that completely fill part of their outer shell, which makes it difficult to form ions.

nonbiodegradable plastics Plastics that do not decay and that therefore last for a number of years.

nonelectrolyte Substances that do not conduct electricity when molten or in solution. (Substances such as mercury that conduct electricity but remain unchanged are also nonelectrolytes.)

nonmetal Generally, nonmetals are electronegative and are poor conductors of heat and electricity. They do not have a luster and are not ductile or malleable. Their oxides are acidic.

nonmetal oxide *See* acidic oxide.

normal salt A salt in which metal ions (or other cations) have replaced all the acidic hydrogen atoms in an acid.

normal solution A solution in which one gram equivalent of the substance is dissolved in one liter of solution.

Np Symbol for the element neptunium.

nuclear reaction A reaction affecting the nucleus of an atom that can split (*see* fission) or decay and emit either a particle or radiation. If a beta particle is emitted, the atomic number of the nucleus increases by one. If an alpha particle is emitted, the atomic number decreases by two and the relative atomic mass decreases by four.

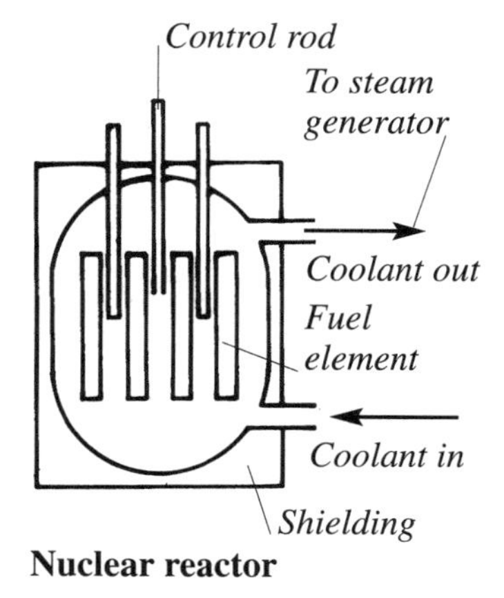

Nuclear reactor

nucleon A proton or neutron.

nucleon number or **mass number** The number of nucleons (protons and neutrons) in the nucleus of an atom.

nucleus The small (about 10^{-14} m diameter) core of an atom. All nuclei contain the positively charged proton, and all but hydrogen contain the zero-charged neutron. The sum of protons and neutrons is the atom's mass number (or nucleon number). The nucleus is surrounded by a cloud of electrons whose number is equal to the number of protons in the nucleus.

nuclide A particular isotope of an element, identified by the number of protons and neutrons in the nucleus.

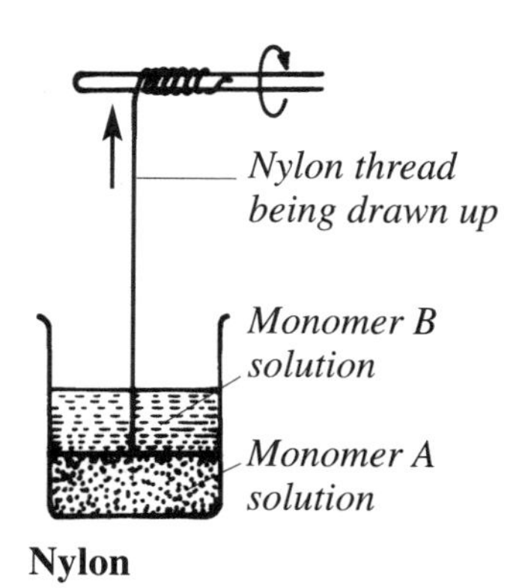

Nylon

nylon A group of polyamides formed by the condensation between an amino group of one molecule and a carboxylic acid group of another. There are three main types of nylon: nylon 6, nylon 6,6, and nylon 6,10. Nylon is very strong, does not rot, and does not absorb water.

O Symbol for the element oxygen.

octane C_8H_{18}. An alkane. A flammable liquid. It has 18 isomers.

octane rating or **octane number** A measure of the quality of gasoline (*see* antiknock). Better fuel contains a higher proportion of molecules with branched chains and has a higher octane rating.

octet A group of eight electrons in the outermost shell of an atom. The noble gases have this structure. In compounds, atoms share (or donate or accept) electrons to form bonds to achieve the octet.

oil–formation *See* fossil fuels.

oil of vitriol Concentrated sulfuric acid.

oil (petroleum)—refining *See* fractional distillation of oil.

oils The general name for mixtures of glycerides with a melting point at room temperature.

olefins *See* alkenes.

oleic acid $C_{17}H_{33}COOH$. A liquid unsaturated fatty acid found in many fats and oils. It is one of the fatty acids used in soap manufacture.

oleum $H_2S_2O_7$ (fuming sulfuric acid). A solution of sulfur(VI) oxide (SO_3) in concentrated sulfuric acid.

orbital An area around an atom or molecule where there is a high probability of finding an electron. An orbital has a fixed energy level. There are different types of orbitals with different shapes: s orbitals, p orbitals, d orbitals, f orbitals, etc. Each orbital can hold two electrons. Orbitals are grouped in a series of shells at a gradually increasing distance from the nucleus. *See* electronic structure of atom.

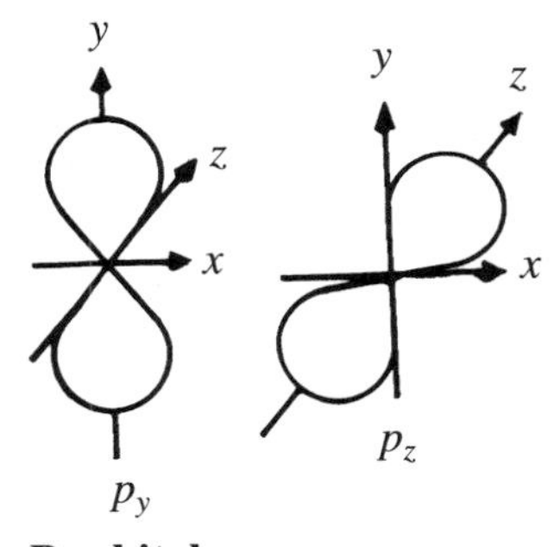

P orbitals

orbit of electrons *See* orbital.

order of reaction A first-order reaction is one in which there is a spontaneous decomposition of one molecule; one that takes place with two molecules is a second-order reaction, and so on.

ore A mineral from which a metal or nonmetal may be profitably extracted.

organic Relates to either living organisms or compounds containing carbon (except carbonates, hydrogen carbonates, and carbon dioxide).

organic acid A group of acids whose structure includes the carboxyl group. Their general formula is $C_n H_{(2n+1)} COOH$.

organic compounds Compounds containing carbon but not carbonates or carbon dioxide.

Os Symbol for the element osmium.

osmium Element symbol, Os; transition element; blue-white metal; Z 76; A(r) 190.2; density (at 20°C), 22.5 g/cm^3; m.p., 3045°C; name derived from the Greek *osme*, "smell"; discovered 1803; used in alloys.

osmosis The movement of solvent molecules through a semipermeable membrane from a dilute solution to a more concentrated solution. There is a tendency for solutions separated in this way to become equal in concentration, and osmosis will stop when equilibrium is

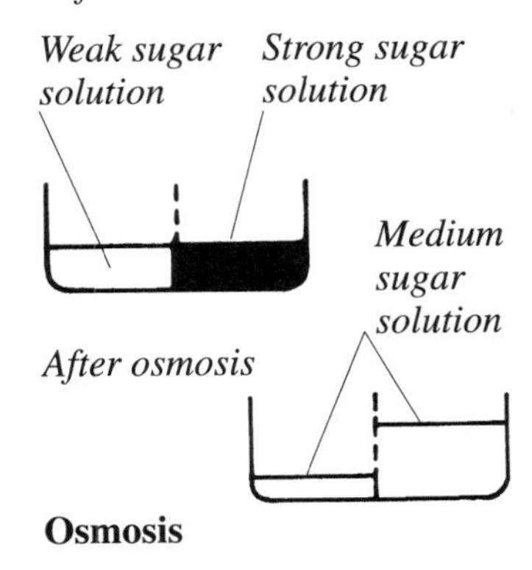

Osmosis

reached. Osmosis can also stop if pressure is applied to the stronger solution (*see* osmotic pressure).

osmotic pressure The pressure that must be applied to a solution, when separated from a more dilute solution by a semipermeable membrane, to prevent the inflow of solvent molecules.

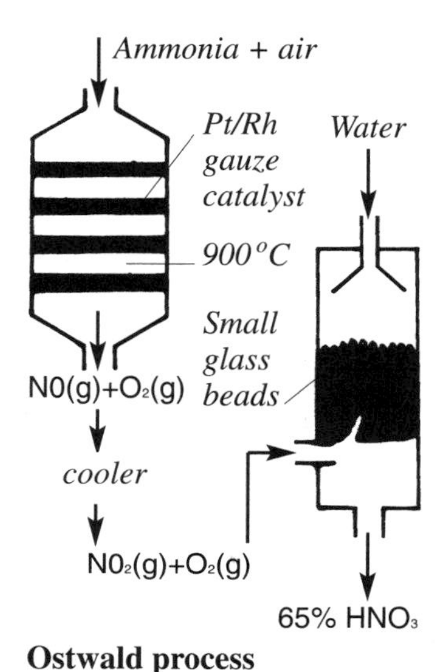

Ostwald process

Ostwald process The manufacture of nitric acid by the catalytic oxidation of ammonia. In the first step of the process, compressed air and ammonia react (at 800°C in the presence of a platinum gauze catalyst) to give nitrogen monoxide and water.

$4NH_3 + 5O_2 \rightarrow 4NO + 6H_2O$.

The nitrogen monoxide cools rapidly and reacts with additional oxygen to form nitrogen dioxide $2NO + O_2 \rightarrow 2NO_2$.

The nitrogen dioxide is cooled, mixed with more oxygen, and passed through water, forming nitric acid. $4NO_2 + O_2 + 2H_2O \rightarrow 4HNO_3$.

oxidation A substance is oxidized if it gains oxygen, loses hydrogen, or loses electrons.

oxidation number *See* oxidation state.

oxidation state This gives an indication of the electron control that an atom has in a compound compared with that which it has in a pure element.

It has two parts. One is the sign: if control has increased, it is negative; if it has decreased, positive. The other part is the value, which gives the difference between the number of electrons controlled by the atom in the element and by the atom in a compound.

In oxidation there is an increase in oxidation number. When naming compounds, the oxidation state is given in Roman numerals.

oxide A compound consisting of oxygen and another element only. They can be either ionic or covalent, and there are four types of oxide – acidic, basic, neutral, and amphoteric.

oxidizing agent A substance that can cause the oxidation of another substance by being reduced itself.

oxonium ion H_3O^+. The hydrated hydrogen ion formed from the combination of a hydrogen ion (single proton) with a water molecule.

oxygen Element symbol, O; group 6; colorless, odorless gas; Z 8; A(r) 15.9994; density (at 20°C), 1.429 g/l; m.p., –218.4°C; commonly exists as diatomic form (O_2) but also forms the allotrope ozone (O_3); name derived from the Greek words *oxys*, “acid,” and *genes*, “producing”; discovered 1774; used medically in breathing apparatus and has many industrial uses. Also used in rocket fuels.

oxygen—atomicity of *See* atomicity.

ozone One of the two allotropes of oxygen, existing as O_3. It is a bluish gas with a penetrating smell. It is a very strong oxidizing agent.

P Symbol for the element phosphorus.

Pa Symbol for the element protoactinium.

palladium Element symbol, Pd; transition element; soft white ductile metal; Z 46; A(r) 106.42; density (at 20°C), 12.16 g/cm^3; m.p., 1554°C; named after the asteroid Pallas; discovered 1803; used in dental work and as catalyst in hydrogenation.

palmitic acid $C_{15}H_{31}COOH$. A solid saturated fatty acid found in many fats and oils. It is one of the fatty acids used in soap manufacture.

paper chromatography A way in which some substances can be separated and identified. A spot of the mixture being investigated is placed at one edge of a piece of paper suspended in a solvent. The spot separates into its components and the components move up the paper at different rates, depending on their affinity for the paper and for the solvent used. When the paper is removed and dried, the different components appear as a line of spots along the paper and they can be identified by the distance they have traveled in a measured time.

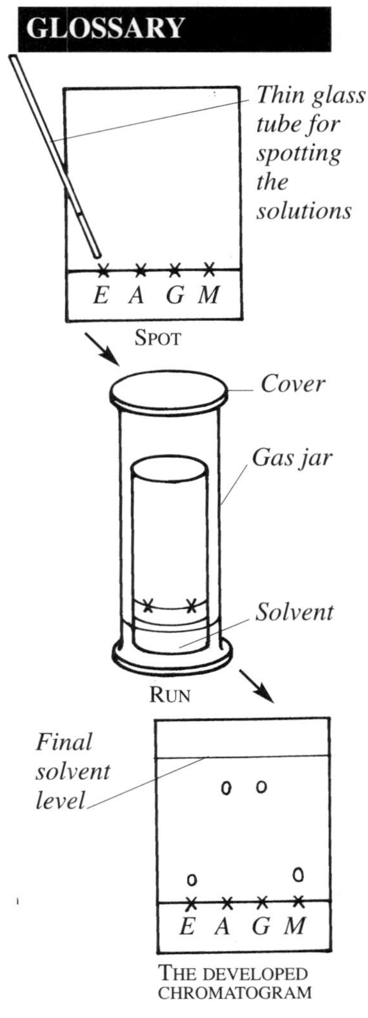

Paper chromatography

paraffins Former name for alkanes, meaning "little affinity."

partial pressure *See* Dalton's law of partial pressure.

passive A metal that is unreactive because its surface is covered with a layer of oxide.

Pb Symbol for the element lead.

Pd Symbol for the element palladium.

pentane C_5H_{12}. An alkane; m.p., –129.7°C; b.p., 36.1°C.

peptide An organic substance consisting of two or more amino acid units joined by peptide bonds. The bonds are formed by a condensation reaction between the carboxyl group of one amino acid and the amino group of another. A molecule of water is eliminated as a peptide bond is formed.

peptide bond A link joining amino acid units, forming peptides. One end of the link is a carbon atom that has a double bond to an oxygen atom and a single bond to a nitrogen atom. The nitrogen atom forms the other end of the link and also has a single bond to a hydrogen atom.

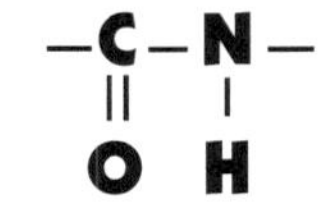

Peptide bond

percentage composition The proportion by mass of the component parts of a compound expressed as a percentage of the mass of the whole compound. When the percentage composition of a substance and the relative atomic mass of each element are known, the empirical formula of the compound can be calculated.

period The horizontal rows of elements in the periodic table.

The first three rows (hydrogen–helium, two elements; lithium–neon, eight elements; sodium– argon, eight elements) are short periods.

The next four rows include the elements known as transition elements (potassium–krypton, 18 elements; rubidium–xenon, 18 elements; cesium–radon, 32 elements; francium and above, 26 elements discovered so far) and are therefore long periods. The cesium–radon period also includes the lanthanides, and the period beginning with francium includes the actinides.

The atoms of the elements in a period have the same number of shells. The number of electrons in the outer shell increases by one for each position moved to the right in the periodic table. There is a change in behavior of the elements in a period from metallic (electropositive) on the left of the periodic table, to nonmetallic (electronegative) at the right of the periodic table.

periodic law Proposed in 1869 by Russian chemist Dmitri Mendeleyev, it is the basis of the modern periodic table of the elements. On arranging the elements known at that time in ascending order of their relative atomic masses, he discovered that elements with similar chemical properties appeared at fixed intervals, or periods. There were gaps in this series of elements, and these led him to predict that elements of a certain relative atomic mass would be discovered to have certain physical and chemical properties.

periodic table A table of elements arranged in ascending order of atomic number. It has eight main groups (*see* group) and seven periods (*see* period). Knowing an element's position in the periodic table enables its physical and chemical properties to be predicted.

permanent hardness Calcium or magnesium sulfates that react with the sodium stearate molecules in soap to form a scum on the surface of the water. *See* soap, hardness in water.

Permutit A compound that can soften water. It does this by exchanging its sodium ions for the calcium and magnesium ions in hard water. It consists of sodium aluminum silicate. It is a zeolite. The sodium ions do not form insoluble salts with soap and therefore do not prevent the formation of a lather. The Permutit can be returned to its original state by soaking in brine.

peroxide A compound that contains the peroxide ion O_2^{2-}. Peroxides are strong oxidizing agents. Hydrogen peroxide H_2O_2 (produced by the action of water or dilute acids on sodium peroxide) is used in dilute form as a bleach and disinfectant.

perspective formula Representation of a molecule on paper: a solid line shows a bond in the same plane as the paper; a dotted line, a bond behind the paper; a wedge shape, a bond pointing outwards.

Perspex *See* poly(methylmethacrylate).

petrochemical A chemical made from petroleum.

petrol *See* gasoline.

pH A scale that gives a measure of the acidity of an aqueous solution. The concentration of hydrogen ions is used in the calculations, and the pH value of a solution is given as $\log_{10}(1/H^+)$, where H^+ is the concentration of hydrogen ions. A neutral solution has a pH of 7, while an acidic solution has a lower value and an alkaline solution a higher value.

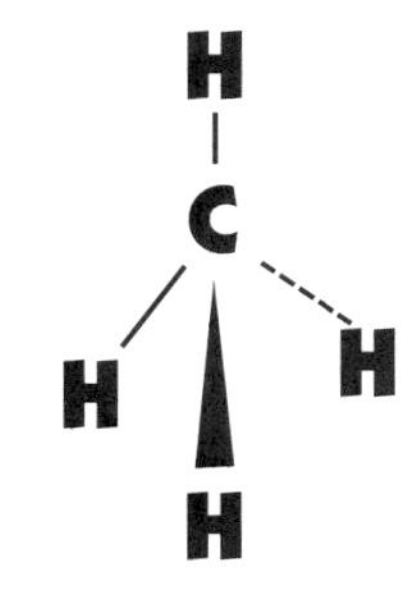

Perspective formula for methane

phase Part of a system whose physical properties and chemical composition are consistent and are separated from other parts of the system by a boundary surface. For example, two immiscible liquids form a two-phase liquid system; a vessel containing ice, water, and water vapor is a three-phase system.

phase change The change that occurs when a substance changes its physical state, between being a solid, liquid, or gas or being in solution.

phase diagram A diagram showing the change between states for a substance at different conditions of pressure and temperature.

phenol C_6H_5OH. An aromatic organic compound; m.p., 43°C; b.p., 183°C. It is an acidic, poisonous, corrosive crystalline compound that forms metallic salts. Phenol is colorless, but turns pink on exposure to air and light. It is soluble in water at room temperatures; its solution is called carbolic acid, which is used as a disinfectant. Phenol can be obtained from coal tar, and it is readily halogenated, sulphonated, and nitrated. Phenol is used in the manufacture of phenol/methanal resins, poly(carbonates), epoxy resins, nylon, dyes, and detergents.

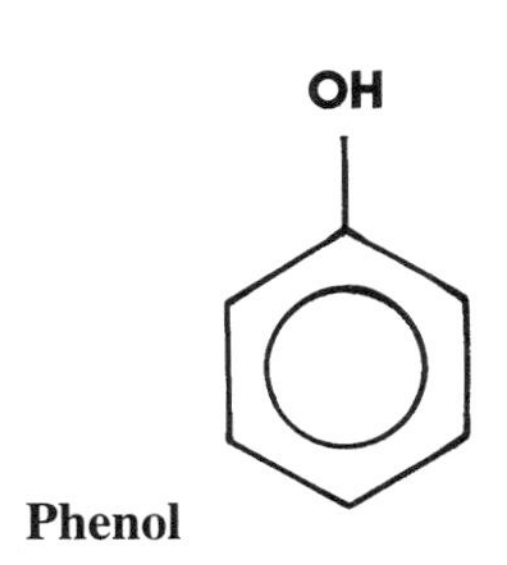

Phenol

phenol/methanal resins Polymers made by a condensation reaction between phenol and methanal. They are dark in color and are good electrical insulators. Bakelite is a phenol/methanal resin.

phenolphthalein An acid-base indicator used in titrations of a weak acid with a strong alkali. Its colorless crystals turn pink when the solution is alkaline.

phenylethene $C_6H_5CH{=}CH_2$. A liquid aromatic hydrocarbon; m.p., –31°C; b.p., 145°C.

phenyl group The organic group C_6H_5– present in benzene.

phosphates Salts of phosphoric acid H_3PO_4. As this is a tribasic acid, three types of phosphate can be formed: the alkaline phosphate (containing the trivalent radical PO_4), the neutral hydrogenphosphate (containing the divalent radical HPO_4), and the acidic dihydrogephosphate (containing the monovalent radical H_2PO_4).

phosphorescence The emission of light by an object, and the persistence of this emission over long periods, following irradiation by light or other forms of radiation. Energy is absorbed by the object and then re-radiated at a longer wavelength than the incident light. White phosphorus, zinc sulfide, and calcium sulfide are phosphorescent substances.

phosphoric(V) acid H_3PO_4 (orthophosphoric acid). A white, very deliquescent crystalline solid; m.p., 42.35°C. It is very soluble in water, forming a weak tribasic acid. It is also soluble in ethanol. Commercially produced from phosphate-containing rocks. *See* phosphates.

phosphorus Element symbol, P; group 5; three main allotropes: white (containing tetrahedral P4 atoms), red (a polymer), and black (structure like graphite); Z 15; A(r) 30.97; density (at 20°C), 1.82 g/cm^3 (white); m.p., 44.1°C (white); name derived from the Greek *phosphorus*, "light-bearing"; discovered 1839; used in fertilizers and matches.

phosphorus chlorides PCl_3 (phosphorus(III) chloride, phosphorus trichloride). A colorless fuming liquid, m.p., –112°C; b.p., 75.5°C. It hydrolyzes violently to form phosphonic acid (H_3PO_3), which is used in organic synthesis to replace an –OH group with a chlorine atom.

PCl_5 (phosphorus(V) chloride, phosphorus pentachloride). A yellow-white crystalline solid that fumes in air; m.p., 148°C (under pressure). It sublimes between 160–162°C and decomposes in water to form phosphoric acid and hydrogen chloride. It is also used as a chlorinating agent.

phosphorus oxides P_2O_3 (phosphorus(III) oxide, phosphorus trioxide). A white, or colorless, waxy solid; m.p., 23.8°C; b.p. 173.8°C; reacts with cold water, forming phosphonic acid (H_3PO_3). With hot water it reacts to form phosphine gas (PH_3) and phosphoric acid.

P_2O_5 (phosphorus(V) oxide, phosphorus pentoxide); m.p., 580°C under pressure. It sublimes at 360°C and reacts violently with water, forming phosphoric acid. It is used as a drying agent and as a dehydrating agent.

photocatalytic The speeding up or slowing of a chemical reaction by light.

photochemical reaction A chemical reaction that is initiated by a particular wavelength of light.

photolysis The decomposition or disassociation of a compound when exposed to light of a certain wavelength.

photosynthesis This is an important photochemical reaction. It is the process by which green plants make carbohydrates using carbon dioxide and water. Oxygen is also produced.

phototropy The ability of certain substances to change color reversibly on exposure to light of a certain wavelength.

physical change A reversible change, such as a change of state, where no chemical reaction takes place and no new substances are formed. A reversible color change is also a physical change.

physical chemistry The branch of chemistry concerned with the study of the physical properties of elements and compounds, and the relationship between these properties and their chemical structure.

pigment An organic or inorganic chemical that has a characteristic color.

pi orbital *See* molecular orbital.

pipette A glass tube that is used to measure and transfer a fixed volume of liquid. Pipettes are available in a range of volumes. Suction is applied to the top of a pipette to draw a liquid up so that its meniscus is on the marked line on the pipette. When the suction is released, the liquid flows out of the pipette.

planar A molecule whose atoms are in the same plane. (It is flat.)

plaster of Paris *See* calcium sulfate.

plastic A substance that can be shaped by heating and pressure during manufacture to form a stable product.

plasticizer A substance added to polymers and other materials to increase their flexibility.

plastics—burning When plastics burn, they emit poisonous gases. The type of gas depends on the particular plastic, but all plastics will emit carbon monoxide if combustion is incomplete. Polyurethane plastics contain nitrogen, which combines with the carbon and hydrogen that are also in the plastic to form hydrogen cyanide. Hydrogen chloride is produced from the combustion of polychloroethene.

platinum Element symbol, Pt; transition element; soft, shiny silver metal that is malleable and ductile; Z 78; A(r) 195.09; density (at 20°C), 21.4 g/cm^3; m.p., 1772°C; name derived from the Spanish *plata*, "silver"; discovered 1735; used in jewelry and electrical contacts; also used as a catalyst in many processes, including the removal of harmful substances from vehicle exhaust gases.

plutonium Element symbol, Pu; actinide; silvery metal, very radioactive; Z 94; A(r) 244; density (at 20°C), 19.8 g/cm^3; m.p., 641°C; 13 isotopes known; named after the planet Pluto; discovered 1940; plutonium-238 is used in nuclear reactors as a power source; plutonium-239 is used in nuclear weapons and some nuclear reactors.

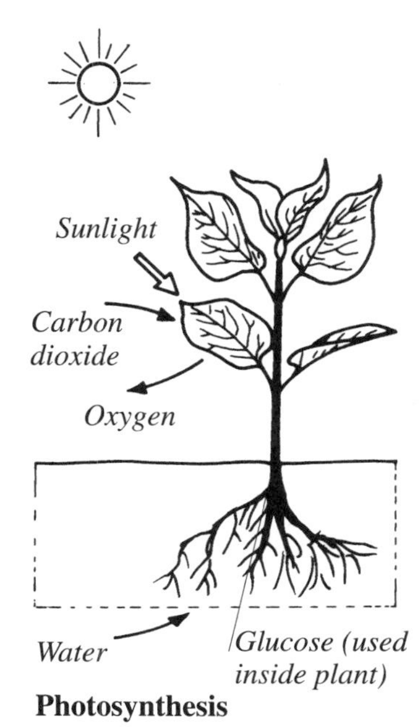

Photosynthesis

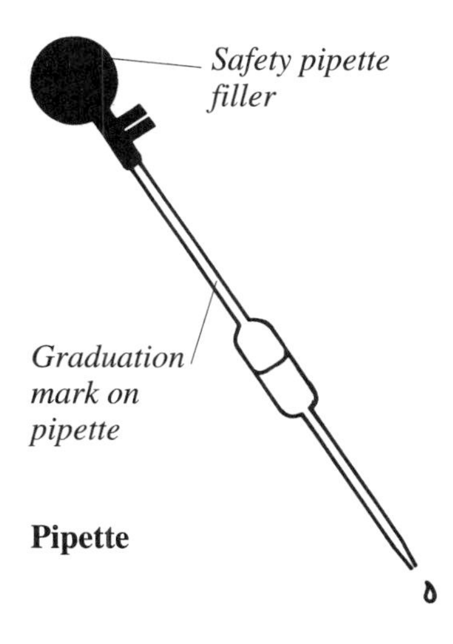

Pipette

Pm Symbol for the element promethium.

Po Symbol for the element polonium.

poison (1) A substance that causes damage to a living organism.
(2) A substance that destroys the activity of a catalyst.

polar molecule A molecule that has a positive charge at one end and a negative charge at the other. This occurs because the two electrons in covalent bonds are not shared equally between atoms that have different electronegativities. This leads to a separation of charge across the bond. If the effects are not canceled out over the molecule as a whole, the molecule becomes polar. Lone pairs of electrons also cause a molecule to be strongly polar. *See* molecules – shapes of, core charge.

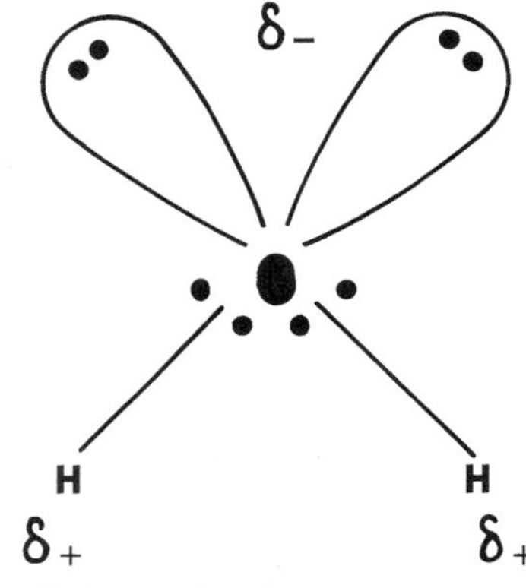

Polar molecule

pollution Harmful contamination of the environment caused by, for example, poor disposal of waste products, including exhaust gases, escape of dangerous substances (such as leaks from petrochemical tankers), and excessive amounts of nitrate-containing fertilizers being washed into rivers and lakes.

polonium Element symbol, Po; group 6; radioactive metal; Z 84; A(r) 210; density (at 20°C), 9.4 g/cm^3; m.p., 254°C; name derived from Medieval Latin *Polonia*, "Poland"; discovered 1898.

poly- Prefix meaning "many," used in the naming of chemical compounds.

polyamide A condensation polymer that contains the amide group. Nylon is a polyamide.

polybasic acid An acid that has more than one hydrogen atom that can be replaced to form a salt.

polycarbonates Thermoplastics made by condensation polymerization of carbonyl chloride (phosgene) and dihydroxy organic compounds such as diphenylol propane. Polycarbonates are tough and transparent; they are used for spectacle lenses, babies' bottles, and shatterproof windows.

polychloroethene (polyvinyl chloride, PVC) Thermoplastic polymer made by addition polymerization from chloroethene. It is a very tough white solid material and is easy to color. It is resistant to fire, chemicals, and weather, and has many uses. It is used as a floor covering, for artificial leathers, containers, and drainage pipes.

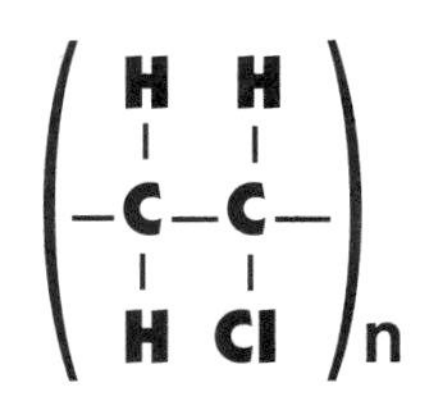

Polychloroethene

polycondensation *See* condensation polymerization.

polyesters A group of condensation polymers formed from a polybasic carboxylic acid and a polyhydric alcohol. They contain ester groups.

polyethene (polythene, polyethylene). Thermoplastic polymer made by addition polymerization of ethene. Polyethene is a saturated alkane and is thus very unreactive. Polyethene is a tough white waxy solid that is unaffected by acids, bases, or solvents, or oxidizing or reducing agents. It is flexible and a good insulator. It can be manufactured in one way to produce low-density polyethene or LDPE. This is very flexible and can be manufactured in sheets for packaging. The high-density form (HDPE) can also be made. This is also flexible and can be blow-molded to produce containers. Both types can be used for injection molding to manufacture boxes and bowls, etc.

polymer A material containing very large molecules that are built up from a series of small basic units (monomers). It is a term often applied to plastics. There can be between hundreds and hundreds of thousands of basic units in a polymer.

polymerization The formation of a polymer from monomers. There are two types of polymerization reactions: addition and condensation. Polymers formed from a single monomer are called homopolymers. Polymers formed from two or more monomers are called copolymers.

$$nC_3H_6 \rightarrow (C_3H_6)n$$

Polymerization of propene

polymethylmethacrylate (polymethyl 2-methylpropenoate, Perspex) A transparent thermoplastic addition polymer. Made by polymerizing methyl methacrylate (methyl 2-methylpropenoate), it contains many ester groups and is thus called a polyester. It is lighter and stronger than glass but more easily scratched. It is used for airplane windows and car lights.

polymorphism A substance's potential to exist in more than one form. Allotropy is one form of polymorphism, but polymorphism also covers noncrystalline forms. Each polymorphic form of a substance is stable within a range of physical conditions (temperature, pressure) and will transform to another polymorphic form at a fixed transition temperature.

polypeptide A peptide that contains at least 10 amino acids. Protein molecules usually contain between 100 and 300 amino acids. The particular amino acids present in the structure of the polypeptide and the sequence in which they occur determine its properties. Enzymes are polypeptides.

polyphenylethene (polystyrene) A polymer made by the addition polymerization of phenylethene. Polyphenylethene is resistant to water, acids, alkalis, and solvents. It is similar to polyethene and can be used as a glass substitute. It is often seen in its expanded form – after air or carbon dioxide has been blown into it – when it forms an opaque solid, which has good insulating properties. In this form it is used for cups and packaging material in the fast-food sector.

polypropene (polypropylene) Thermoplastic polymer made by addition polymerization of propene. It is similar to polyethene but with greater resistance to heat and organic solvents. It is strong and hard-wearing, is used for carpet, and injection molded for car fenders.

polysaccharides *See* carbohydrate.

polystyrene *See* polyphenylethene.

polytetrafluoroethene (PTFE, Teflon, Fluon) Thermosetting polymer formed by the polymerization of polytetrafluoroethene. A very inert substance that is very resistant to chemicals, heat, and wear. It is used to give a nonstick coating to cooking utensils. It has a low coefficient of friction and is used for bearings and in replacement joints in the body.

polythene *See* polyethene.

polyurethanes A wide range of condensation polymers that can be either thermosetting or thermoplastic. They contain the urethane (–NH.CO.O–) group. They are formed from polyhydric alcohols and organic isocyanates. They can be found in adhesives, paints, and plastics. If water is added during manufacture, polyurethanes form into a foam, which can be either rigid or flexible. These foams are used in upholstery, insulation, and carpet backing.

polyvinyl chloride *See* polychloroethene.

p orbital A type of orbital. Three types are possible, each of which can hold two electrons. *See* molecular orbital.

porous Able to allow the passage of water, air, or other fluids.

post-actinide elements *See* transactinide elements.

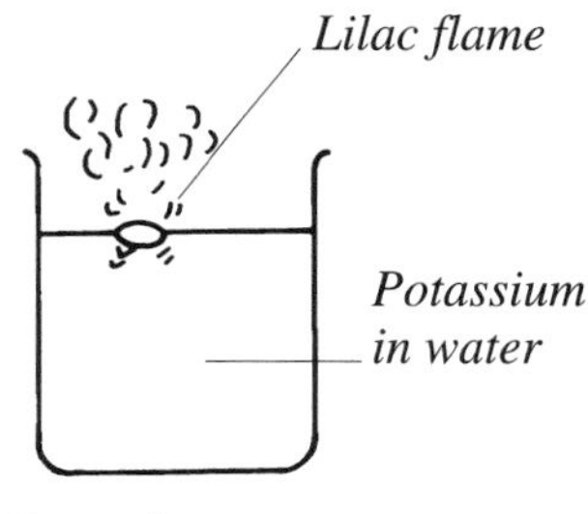

Potassium

potassium Element symbol, K; alkali metal, group 1; soft silver-white metal; Z 19; A(r) 39.1; density (at 20°C), 0.87 g/cm^3; m.p., 63.3°C; reacts violently with water; used as reducing agent; name derived from English *potash*; symbol name derived from modern Latin *kalium*; alkali; discovered 1807; used in fertilizers.

potassium bromide KBr. An ionic compound that exists as $K^+ Br^-$ (m.p., 750°C). It is nonvolatile and soluble in water. Its aqueous solution is an electrolyte.

potassium carbonate K_2CO_3. A white deliquescent solid (m.p., 891°C) that is soluble in water, forming an alkaline solution. It is used as a drying agent in the manufacture of soft soap and in the manufacture of hard glass.

potassium chlorate $KClO_3$. A white solid soluble salt (m.p., 360°C) that decomposes above 400°C, giving off oxygen. It is a powerful oxidizing agent. Potassium chlorate is used in matches, fireworks, explosives, weed killers, etc.

potassium chloride KCl. An ionic compound that exists as $K^+ Cl^-$ (m.p., 790°C). It is nonvolatile and is soluble in water. Its aqueous solution is an electrolyte.

potassium hydrogencarbonate $KHCO_3$ (potassium bicarbonate) A white crystalline solid that is soluble in water. It decomposes at about 120°C. It is used in baking, soft drinks, and carbon dioxide fire extinguishers. A solution of potassium hydrogencarbonate makes a good buffer solution.

potassium hydrogensulfate $KHSO_4$ (potassium bisulfate).

potassium hydroxide KOH. A white deliquescent solid (m.p., 306°C; b.p., 1320°C). It is soluble in water, the aqueous solution is a strong alkali. It is used in the manufacturing of soap and fertilizers, as an electrolyte in batteries, and to absorb acidic gases such as carbon dioxide and sulfur dioxide.

potassium iodide KI. A white crystalline solid; m.p., 686°C; b.p., 1330°C. It is prepared by adding iodine to a hot concentrated aqueous solution of potassium hydroxide and separating the resulting potassium iodide from the potassium iodate that is also produced. It is soluble in water. Potassium iodide is used medically in the treatment of iodine deficiency diseases.

potassium nitrate KNO_3 (niter, saltpeter) A white crystalline solid that is soluble in water; m.p., 334°C; decomposes at 400°C. It is prepared by the double decomposition of boiling saturated solutions of sodium nitrate and potassium chloride, followed by fractional crystallization (sodium chloride crystallizes out first at the temperature of the reaction). Potassium nitrate is a strong oxidizing agent, and it is used in gunpowder, fertilizer, and fireworks.

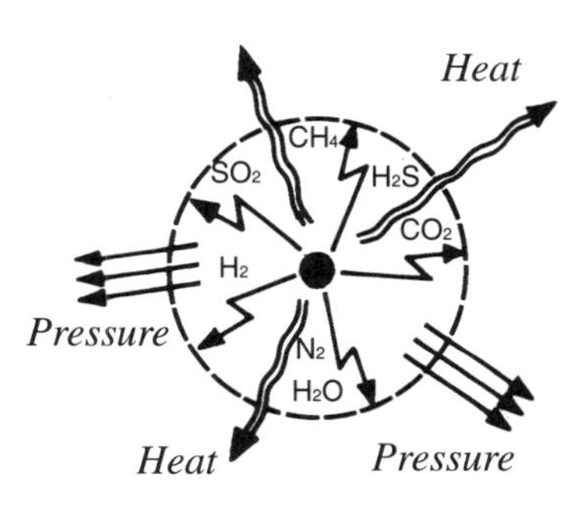

Potassium nitrate acts as an oxidizing agent, producing gases.

Potassium nitrate – gunpower

potassium silicate K_2SiO_3. It is used, with calcium silicate, in the manufacture of hard glass, which has a higher melting point than ordinary (soda) glass, a mixture of sodium and calcium silicates.

potassium sulfate K_2SO_4. A white crystalline solid; m.p., 1072°C. It is soluble in water and can be prepared by neutralizing potassium hydroxide with sulfuric acid. It is found as the mineral schonite, and it is used in fertilizers, cements, and glass.

Pr Symbol for the element praseodymium.

praseodymium Element symbol, Pr; rare earth /lanthanide; soft silvery metal; Z 59; A(r) 140.91; density (at 20 C), 6.77 (white) g/cm^3; m.p., 931°C;

name derived from the Greek words *prasios*, "green," and *didymos*, "twin"; discovered 1885.

precipitate An insoluble substance formed by a chemical reaction. Precipitation is the process by which a precipitate is formed.

principle of conservation of energy Energy is neither created nor destroyed in a chemical reaction.

producer gas Producer gas is a mixture of one part of carbon monoxide to two parts of nitrogen. It is formed by blowing air through hot coke in a "gas producer." It is much cheaper than coal gas but has a lower calorific value.

$C + O_2 \rightarrow CO_2$ + heat.

$CO_2 + C \rightarrow 2CO$ – heat.

Or 2C+ air ($O_2 + 4N_2$) $\rightarrow 2CO + 4N_2$ + heat.

product A substance produced during a chemical reaction. In an equation describing a chemical reaction, the products are shown to the right of the arrow.

promethium Element symbol, Pm; rare earth/lanthanide; soft silvery metal; Z 61; A(r) 145; density (at 20°C), 7.26 g/cm^3; m.p., 1168°C; named after Promethius, mythical Greek character; discovered 1945; used in luminous paint for watches.

propane C_3H_8. An alkane. It is a colorless flammable gas (m.p., –189°C; b.p., –42°C) that is found in natural gas and petroleum. It is used as a fuel and in the synthesis of organic compounds. When liquefied it is known as liquefied petroleum gas (LPG) and is a clean-burning fuel.

propanone CH_3COOCH_3 (acetone) A ketone; m.p., –95.4°C; b.p., 56.2°C.

propene C_2H_4 (propylene) An alkene. It is a colorless gas (m.p., –81°C; b.p., 48.8°C) that is made by cracking petroleum. It is used in the manufacture of polypropene and other organic chemicals such as propanone and glycerin.

properties The intrinsic features of a substance that can identify it. Physical properties include features such as color, boiling and melting points, crystal form, and solubility. Chemical properties include identifying if the substance is a metal or nonmetal, an oxidizing or reducing agent, its valency, and the result of a reaction with acid.

propyl group The organic group $–C_3H_7$.

proteins Proteins are important in the nutrition, structure, and function of living organisms. They are large polypeptides, and the particular amino acids present in the structure of the protein and the sequence in which they occur determine their properties.

Proteins found in skin, hair, and muscle are fibrous proteins; they are insoluble in water.

Enzymes and protein hormones (such as insulin) are globular proteins, which are soluble in water.

Bond formation in a protein determines its structure. The helical shape of fibrous proteins is caused by hydrogen bonding between N–H and C=O groups.

protoactinium Element symbol, Pa; actinide; Z 91; A(r) 231.04; density (at 20°C), 15.37 g/cm^3; m.p., 1200°C; name derived from the Greek word *protos*, "first," and *actinium*; discovered 1917.

proton One of the basic particles in the atom, found in the nucleus with the neutron. It is one of the most massive of the subatomic particles, similar in mass to the neutron. It has positive charge. In a neutral atom the number of protons is equal to the number of electrons. Its mass is 1.673×10^{-27} kg.

protonated Containing an additional proton (or hydrogen ion H^+). For example, the protonated water molecule is the hydroxonium ion H_3O^+.

proton number *See* atomic number.

Proust's law *See* law of constant composition.

Pt Symbol for the element platinum.

PTFE *See* polytetrafluoroethene.

Pu Symbol for the element plutonium.

PVC *See* polychloroethene.

pyridine C_6H_5N. An aromatic heterocyclic compound. It is a very stable colorless liquid with an unpleasant smell. It is used as a solvent.

pyrites A mineral containing metal sulfides such as iron(II) disulfide (FeS_2).

pyrolysis The decomposition of a substance by heat. *See* cracking.

quadrivalent (tetravalent) Having a valency of four.

qualitative A statement, or analysis, that gives the composition of an item, not the amounts present.

qualitative analysis The analysis of a sample of an unknown compound to identify its constituent parts. Such analysis is done using chemical and physical tests (for example, flame test).

quantitative A statement, or analysis, that gives the amounts of an item present.

quantitative analysis The analysis of a sample of a compound whose component parts are known in order to estimate the amounts of the

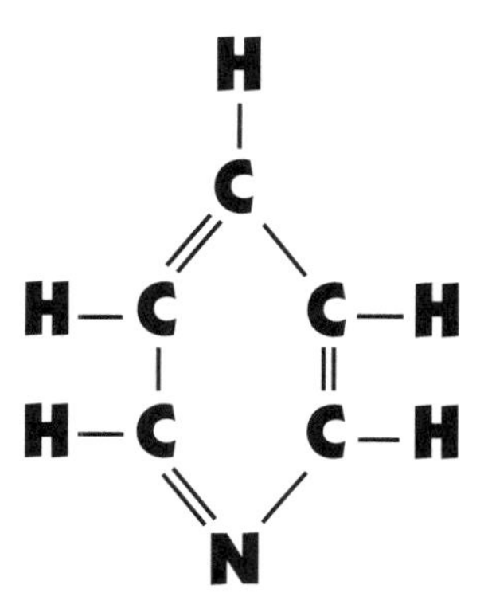

Pyridine heterocyclic compound

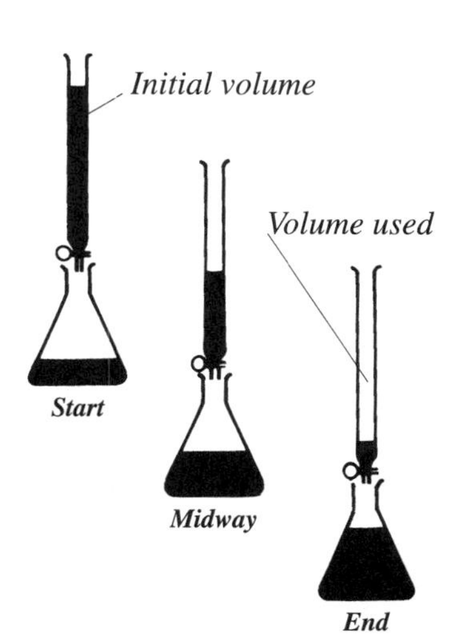

Quantitive analysis

component parts present in the sample. Volumetric and gravimetric methods can be used.

quicklime *See* calcium oxide.

Ra Symbol for the element radium.

radical A group of atoms forming part of many molecules. Radicals are very reactive as they have an incomplete electron structure.

radioactive series There are three naturally occurring radioactive series - the thorium series, the actinium series, and the uranium series. Each series is headed by the named element and radioactive decay of this element proceeds. Alpha and beta particles and gamma radiation are emitted at different stages, creating different nuclides that undergo further decay until a stable nuclide (lead in all three cases) is formed.

radioactive tracers Labeling of non-radioactive material by adding small quantities of a radioactive preparation to study the movement of the material.

radioactivity The spontaneous disintegration of certain isotopes accompanied by the emission of radiation (alpha particles, beta particles, gamma waves).

radiocarbon dating *See* carbon dating.

radium Element symbol, Ra; alkaline earth metal, group 2; radioactive white metal; Z 88; A(r) 226.03; density (at 20°C), 5 g/cm^3; m.p., 700°C; name derived from the Latin *radius*, "ray"; discovered 1898; used in luminous paints, neutron source, and radiotherapy.

radon Element symbol, Rn; noble gas, group 8; colorless radioactive gas; Z 86; A(r) 222; density (at 20°C), 9.96 g/l at STP; m.p., –71°C; name derived from the radium; discovered 1900.

Raoult's law In a solution at a constant temperature, the vapor pressure of the solvent is lowered in proportion to the mole fraction of the solute.

rare earth elements *See* lanthanides.

rate of reaction For a chemical reaction, a measure of either the amount of reactants used or amount of products formed in unit time. It depends on the concentration of the reactants, temperature, catalyst, and pressure.

raw materials Substances used as a starting point in industrial processes. Important raw materials for the chemical industry include air, water, minerals, hydrocarbons, and metallic ores.

rayon An early synthetic fiber made from wood pulp. There are two methods of making rayon: the acetate method and the viscose method. The acetate method uses cellulose ethanoate, which is dissolved in a solvent and extruded into air through very fine nozzles. The solvent evaporates, leaving the filaments of acetate rayon, which can be spun into threads. In the viscose method the wood pulp is dissolved in carbon disulfide and sodium hydroxide. When this liquid is extruded through fine nozzles into a solution of dilute sulfuric acid, cellulose filaments are produced, which can be spun into threads.

Rb Symbol for the element rubidium.

r.d. Abbreviation for relative density.

Re Symbol for the element rhenium.

reactant A substance present at the start of chemical reaction that takes part in the reaction. In an equation describing a chemical reaction, the reactants are shown to the left of the arrow.

reaction A process in which substances react to form new substances. Bonds are broken and reformed in chemical reactions.

reactivity series of metals (activity series of metals) Metallic elements arranged in order of their decreasing chemical reactivity. Hydrogen is included in the series. Metals placed above hydrogen liberate it from water and dilute acids. A metal may displace another metal from the salt of a metal placed below it in the series. Some elements are in different positions in this series from their positions in the electrochemical series.

recycling metals *See* metals recycling.

recycling plastics The collection of waste plastic materials, sorting into different types (labels usually indicate the type of plastic used), and use of the resulting plastics to manufacture new items.

red lead *See* lead oxides.

redox chemistry A process in which one substance is reduced and another is oxidized at the same time.

reducing agent A chemical that can reduce another while being oxidized itself.

reducing sugar A sugar containing an easily oxidized group, such as an aldehyde or ketone group. All monosaccharides and some disaccharides (lactose and maltose) are reducing sugars.

reduction A chemical reaction in which a substance undergoes one of the following changes – a loss of oxygen, a gain of hydrogen, a gain of one or more electrons. It is the reverse of oxidation.

refining The process by which a substance is purified. This can be done by removing impurities, for example, in the extraction of a metal from its ore. It can also describe the separation of a particular substance from a mixture of similar substances, for example, in the production of certain hydrocarbon products by the fractional distillation of petroleum.

reforming The use of a platinum-based catalyst in the conversion of hydrocarbon molecules into other products. It is an important process in the petrochemical industry. The molecule is not changed in size bu in structure. The octane rating (the proportion of branched hydrocarbon chains to straight chains) of a fuel is improved by the reforming process. This process is also used to form aromatic compounds from alkenes.

Copper wire

Regelation

regelation The melting of ice when subjected to pressure and refreezing on removal of that pressure.

relative atomic mass (A(r)) The ratio of the mass of an average atom of an element to 1/12 of the mass of an atom of the carbon-12 isotope. (Mass of an atom of the carbon-12 isotope is taken as 12.)

relative density (r.d.) The ratio of the density of a substance at 20°C divided by the density of water at 4°C. It is also the ratio of the mass of a volume of the substance to the mass of an equal volume of water (both measure at the same temperature). The relative density of a gas can be given relative to dry air or to hydrogen (all measurements at STP) to that of a reference substance (usually water, for liquids or solids). Formerly called specific gravity.

relative formula mass *See* formula mass.

relative molecular mass (M_r) The ratio of the mass of a molecule of the element or compound to 1/12 of the mass of an atom of the carbon-12 isotope. (Mass of an atom of the carbon-12 isotope is taken as 12.)

repeating unit A group of atoms in the structure of a polymer that is repeated many times.

residue The solid remaining after the completion of a chemical process.

reversible reactions A chemical reaction that can proceed in either direction. It does not reach completion but achieves dynamic equilibrium.

Rf Symbol for the element rutherfordium.

Rh Symbol for the element rhodium.

rhenium Element symbol, Re; transition element; silvery gray metal; Z 75; A(r) 186.2; density (at 20°C), 21.02 g/cm^3; m.p., 3180°C; name derived from the Latin name for the river Rhine (*Rhenus*); discovered 1925; used in alloys, thermocouples, and catalysts.

rhodium Element symbol, Rh; transition element; silvery white metal; Z 45; A(r) 102.91; density (at 20°C), 12.44 g/cm^3; m.p., 1966°C; very inert name derived from the Greek *rhodon*, "rose"; discovered 1803–4; used in platinum alloys.

Rn Symbol for the element radon.

rock salt (halite). A mineral form of sodium chloride. It can be extracted by pumping water into underground deposits where it dissolves the rock salt, producing brine that is evaporated to produce the salt. Rock salt is also produced by evaporation of seawater.

room temperature A temperature in the range 15–25°C.

Ru Symbol for the element ruthenium.

rubber An elastic polymer that is a good insulator. It was originally a natural product formed from the milky sap of rubber trees, but most is now synthetic, derived from butadiene.

rubidium Element symbol, Rb; alkali metal, group 1; silvery white, very reactive metal; Z 37; A(r) 85.47; density (at 20°C), 1.53 g/cm^3; m.p., 38.9°C; very reactive; burns spontaneously in air; name derived from the Latin *rubidus*, "red"; discovered 1861; used as a getter and in photocells.

rusting The way in which iron is attacked by air and water to form rust (hydrated iron oxide) on its surface.

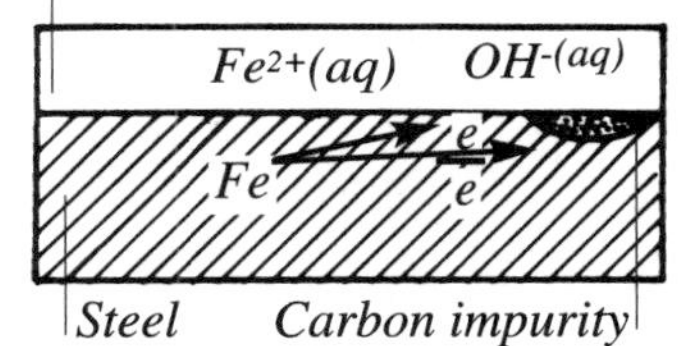

Rusting

ruthenium Element symbol, Ru; transition element; gray white metal; Z 44; A(r) 101.07; density (at 20°C), 12.36 g/cm^3; m.p., 2310°C; name derived from the Latin name *Ruthenia* ("Russia"); discovered 1844.

rutherfordium Element symbol, Rf; transition element; Z 104; A(r) 261; named in honor of Lord Rutherford, the New Zealand physicist; discovered 1964. Formerly known as unnilquadium and kurchatovium.

S Symbol for the element sulfur.

sacrificial anode A sacrificial anode is a block of an electropositive metal in contact with an object being protected from corrosion. It relies on the principle that if two metals are in contact, electrons will flow from the more electropositive metal to the less electropositive metal; the more electropositive metal becomes the anode in the cell created. The sacrificial anode is more electropositive than the object being protected so electrons will flow from the sacrificial anode to the object. In this way steel pipes are protected by magnesium filings and iron hulls of ships are protected by zinc blocks. (*See* sacrificial protection.)

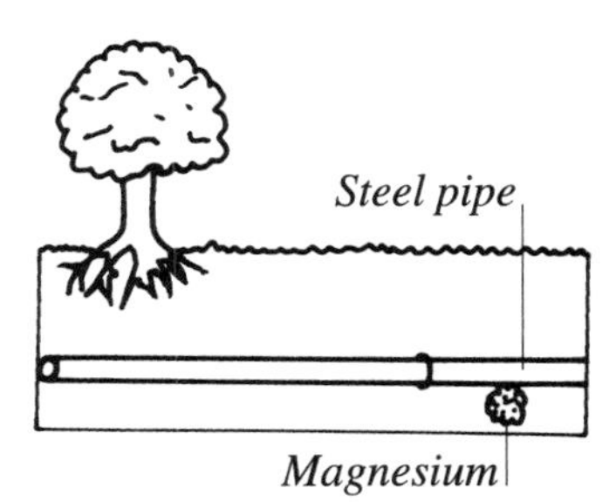

Sacrificial anode

sacrificial protection By attaching a more electropositive metal to the metal that requires protection, the protected metal is no longer corroded because the attached metal has become the anode in the corrosion process and is corroded in its place.

sal ammoniac *See* ammonium chloride.

saline A solution containing one or more salts.

salt A compound formed from an acid in which all or part of the hydrogen atoms are replaced by a metal or metal-like group. They are generally crystalline.

saltpeter *See* potassium nitrate.

salts—preparation of Six common methods:

(1) Action of an acid on a metal.

(2) Action of an acid on an insoluble oxide or hydroxide.

(3) Action of an acid on an insoluble carbonate.

(4) Action of an acid on a soluble base (alkali) or on a soluble carbonate.

(5) Precipitation of an insoluble salt.

(6) Synthesis from its elements.

sal volatile *See* ammonium carbonate.

samarium Element symbol, Sm; rare earth element/lanthanide; soft silvery metal; Z 62; A(r) 150.4; density (at 20°C), 7.52 g/cm^3; m.p., 1077°C; neutron absorber; named after Col. Samarski, Russian mine official; discovered 1879; used to make alloys for nuclear reactor parts; oxide used in permanent magnets.

saponification The treatment of an ester (hydrolysis) with a strong alkaline solution to form a salt of a carboxylic acid and an alcohol. An example is the formation of soap by treating a solution containing esters such as glyceryl stearate with sodium hydroxide to form sodium stearate and the alcohol glycerol.

saturated A solution where there is an equilibrium between the solution and its solute.

saturated hydrocarbon A saturated hydrocarbon that contains only single bonds; it cannot add on extra hydrogen atoms.

saturated solution A solution that can dissolve no more of the solute at a given temperature. There is an equilibrium between the solute and solution.

Sb Symbol for the element antimony.

Sc Symbol for the element scandium.

scandium Element symbol, Sc; transition element; soft, silvery metal; Z 21; A(r) 44.96; density (at 20°C), 2.99 g/cm^3; m.p., 1541°C; name derived from the Latin *Scandia*, “Scandinavia”; discovered 1879; used in small quantities to strengthen alloys.

Se Symbol for the element selenium.

seaborgium Element symbol, Sg; transition element; Z 106; A(r) 263; named in honor of American nuclear chemist Glenn T. Seaborg; discovered 1974. Formerly known as unnilhexium.

sedimentation The settling out of particles in suspension in a liquid at the bottom of the liquid, because of gravity.

seed crystals A small crystal added to a saturated or supersaturated solution to cause crystallization.

selenium Element symbol, Se; group 6; metalloid; several allotropes: red, gray and black; Z 34; A(r) 78.96; density (at 20°C), 4.8 g/cm^3 (gray); m.p., 217°C (gray); element is a semiconductor; name derived from the Greek *selene*, "moon"; discovered 1817; used in electronics; gray allotrope is photosensitive and is used in xerography and photoelectric cells.

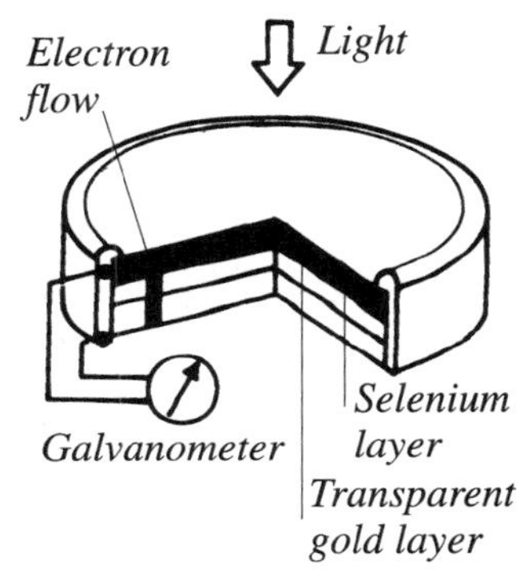

Selenium cell

semipermeable membrane A substance that allows solvent, but not solute, molecules to pass through.

separation of mixtures The separation method used depends on the physical properties of the components of the mixture. Methods include filtration and fractional distillation.

Sg Symbol for the element seaborgium.

shell A group of orbitals that are grouped at a similar distance from an atomic nucleus.

shortened structural formula This gives the sequence of groups of atoms in a molecule, showing which groups of atoms are present, and gives an idea of the molecule's structure, for example, CH_3CH_2OH for ethanol.

CH_3CH_3

Shortened structural formulas

SI Abbreviation for *S*ystème *I*nternational (d'Unités), which proposed a system of coherent metric units (SI units) for international recognition in 1960 that is now the standard system of units used in science.

Si Symbol for the element silicon.

sigma orbital *See* molecular orbital.

silica *See* silicon(IV) oxide.

silica gel Amorphous form of hydrated silica. It is very hygroscopic and is used to absorb water. When saturated it can be regenerated by heat.

silicon Element symbol, Si; group 4; metalloid; Z 14; A(r) 28.09; density (at 20°C), 2.3 g/cm^3; m.p., 1410°C; second most abundant element in Earth's crust (SiO_2); name derived from the Latin *silex*, "flint"; discovered 1823; used in transistors.

silicon dioxide *See* silicon(IV) oxide.

silicon(IV) oxide SiO_2 (silicon dioxide, silica) It is a hard crystalline solid (m.p., 1880°C) occurring naturally as quartz and in sand and flint. In crystals of silicon(IV) oxide, the silicon atoms are bonded tetrahedrally to four oxygen atoms, forming a very rigid structure. Silicon(IV) oxide is used in glass manufacture.

silver Element symbol, Ag; transition element; white shiny ductile metal; Z 47; A(r) 107.87; density (at 20°C), 10.49 g/cm^3; m.p., 961.9°C; name *seolfor* in Old English; element symbol derived from the Latin *argentum*; known since prehistoric times; used in jewelry, electrical components, photography, and as a catalyst.

silver bromide AgBr. An insoluble light yellow salt; m.p., 432°C. Silver bromide dissolves in ammonia solution. On exposure to light it decomposes to form silver and bromine. It is used for photographic emulsions.

silver chloride AgCl. An insoluble white salt, m.p.; 455°C, b.p., 1550°C. It dissolves in ammonia solution. Silver chloride is sensitive to light, slowly decomposing to form silver and chlorine. It is used for photographic emulsions.

silver iodide AgI. An insoluble yellow solid; m.p., 556°C; b.p., 1506°C. It does not dissolve in ammonia solution. Silver iodide is sensitive to light, slowly decomposing to form silver and iodine. It is used for photographic emulsions.

silver mirror test Test for the presence of an aldehyde. The sample to be tested is warmed in a test tube with a quantity of Tollen's reagent. If the sample contains an aldehyde, a bright silver mirror is formed on the inside of the test tube as the complex silver ions $[Ag(NH_3)_2]^+$ in the Tollen's reagent are reduced to silver (ketones do not form a silver mirror in this test).

silver nitrate $AgNO_3$. A very soluble white salt; m.p., 212°C. It decomposes to form silver, oxygen, and nitrogen dioxide on heating. It is used for photographic emulsions. Silver nitrate is used in a test for the presence of chloride ions (*see* chlorides). It is also used to test for bromide and iodide ions.

silver sulfide Ag_2S. A very insoluble black salt. It is precipitated when hydrogen sulfide gas is bubbled through a solution containing silver ions. Argentite is a mineral that contains silver sulfide.

simple formula This shows the ratio of ions present in a compound, for example, $CaCl_2$.

single bond A covalent bond formed by a shared pair of electrons (*see* bond).

slag Waste material that collects on the surface of a molten metal during the process of either extraction or refining. It is composed of oxides, phosphates, silicates, and sulfides.

slaked lime *See* calcium hydroxide.

Sm Symbol for the element samarium.

smelting The process of extracting a metal from its ores. It is usually performed by heating the ore with a flux and a reducing agent.

Sn Symbol for the element tin.

soap A substance that will dissolve grease.

Hard soap is a sodium salt of a long chain fatty acid, such as palmitic acid ($C_{15}H_{31}COOH$), oleic acid ($C_{17}H_{33}COOH$), or stearic acid ($C_{17}H_{35}COOH$).

A soft soap is one where sodium is replaced by potassium.

In the manufacture of soap, a hot concentrated solution of sodium hydroxide is added to vegetable oils or animal fats. Vegetable oils and fats contain many different esters, such as glyceryl stearate (formed from an alcohol such as glycerol and a long-chain fatty acid such as stearic acid). Sodium salts of fatty acids (palmitic, stearic, and oleic) form soap, which separates and floats on the surface when strong brine is added. Glycerol remains at the bottom of the mixture.

Soap dissolves grease because the –COONa end of the sodium stearate molecule is hydrophilic while the hydrocarbon end is hydrophobic and is soluble in the grease of oils and fats. The grease droplets are surrounded by the hydrocarbon ends of the sodium stearate molecule that emulsifies and then splits up the grease.

Unfortunately, the sodium stearate molecules react with any calcium or magnesium salts in hard water and form calcium and magnesium stearates that are insoluble in water and form a scum on the surface of the water.

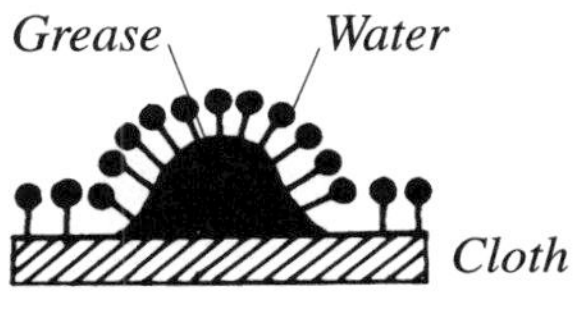

Detergent attacks grease on cloth

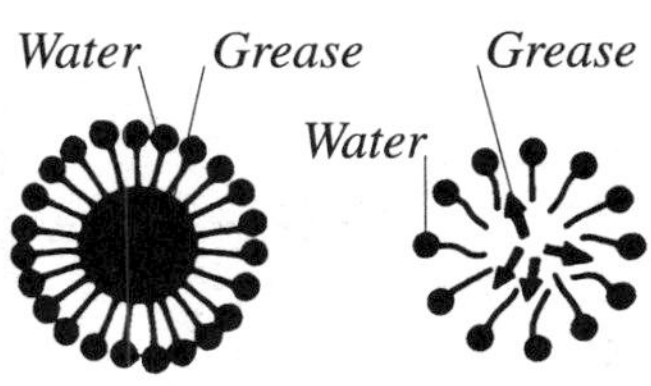

Grease is partly emulsified

Grease is split up into small "soluble" droplets

Soap

soda ash *See* sodium carbonate.

soda-lime A grayish white granular mixture of sodium hydroxide and calcium hydroxide. It is made by adding sodium hydroxide solution to calcium oxide and drying. It is used to absorb carbon dioxide and as a drying agent.

sodium Element symbol, Na; alkali metal, group 1; soft, white, silvery, metal; Z 11; A(r) 22.99; density (at 20°C), 0.97 g/cm^3; m.p., 97.8°C; reacts quickly with water and oxygen. Name derived from the English *soda*; symbol derived from modern Latin *natrium*; discovered 1807; sodium compounds are very important.

soda water A solution of carbon dioxide in water.

sodium acetate *See* sodium ethanoate.

sodium aluminate $NaAlO_2$. A white solid (m.p., 1800°C) that is soluble in water, forming a strong alkali. It is used as a mordant, in the manufacture of glass and zeolites, and in cleaning materials.

sodium bromide NaBr. A white crystalline solid; m.p., 747°C; b.p., 1390°C. It is used medically as a sedative and is also used in analytical chemistry.

sodium carbonate Na_2CO_3. Washing soda ($Na_2CO_3.10H_2O$) is formed on crystallization from an aqueous solution of sodium carbonate. Washing soda is efflorescent; it loses water between 32°C and 34°C to form the monohydrate $Na_2CO_3.H_2O$. This loses water at 109°C. Sodium carbonate is a white solid; m.p., 851°C. It is soluble, forming an alkaline solution. It is manufactured by the Solvay process.

sodium chlorate(I) NaOCl (sodium hypochlorite). Sodium chlorate(I) is soluble, forming an aqueous solution that is used as bleach and as antiseptic.

sodium chlorate(V) $NaClO_3$. A white crystalline solid (m.p., 250°C) that decomposes above 250°C to form oxygen and sodium chloride. It is soluble in water and is a powerful oxidizing agent. Sodium chlorate(V) is used to bleach wood pulp for paper making and is also used as a garden herbicide.

sodium chloride NaCl. (m.p., 803°C) An ionic compound that exists as $Na^+ Cl^-$. It is nonvolatile and is soluble in water. An aqueous solution of sodium chloride is an electrolyte.

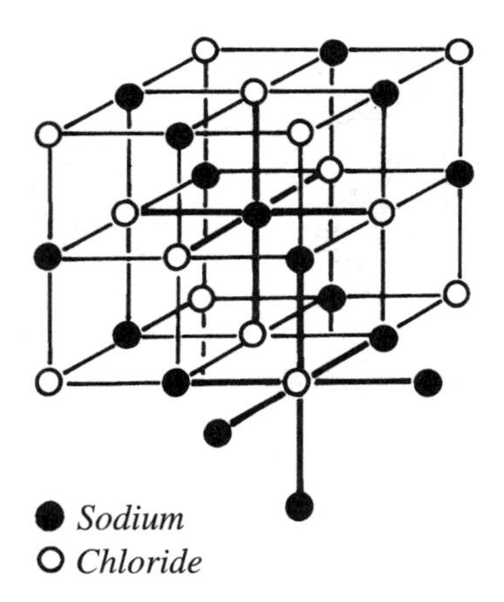

Sodium chloride structure

sodium ethanoate CH_3COONa. A colorless crystalline solid; m.p., 324°C. It is a salt of a strong base and a weak acid and is therefore useful in buffer solutions.

sodium hydrogencarbonate $NaHCO_3$ (sodium bicarbonate, bicarbonate of soda) A white crystalline solid that decomposes at 270°C to form sodium carbonate, carbon dioxide, and water.

$2NaHCO_3 \rightarrow Na_2CO_3 + CO_2 + H_2O$.

It is soluble in water. It is an acid salt but forms an alkaline solution, because HCO_3^- is a stronger base than it is an acid. It is manufactured by the Solvay process. It is used in cooking as baking soda, to neutralize bulk spills of acid, and as an antacid.

sodium hydrogensulfate $NaHSO_4$. A colorless crystalline solid that is soluble in water, forming an acidic solution. It exists in two forms, anhydrous (m.p., above 315°C) and monohydrate, which is deliquescent (m.p., 59°C). Sodium hydrogensulfate decomposes on heating to form sulfur trioxide. It is used in paper and glass manufacture.

sodium hydrogensulfite $NaHSO_3$. A white crystalline solid that turns yellow in aqueous solution. (It is very soluble in water.) It decomposes on heating to form sodium sulfate, sulfur dioxide, and sulfur. It is used in the sterilization of wine casks, as an antiseptic, and as a bleaching agent.

sodium hydroxide NaOH (caustic soda) A white translucent crystalline solid. It is deliquescent and soluble in water, forming a strongly alkaline solution. It is produced by both the Castner-Kellner process and in the diaphragm cell. It is used widely in the laboratory. It is also used in soap manufacture and to absorb acidic gases such as carbon dioxide and sulfur dioxide.

sodium iodide NaI. An ionic compound that exists as $Na^+ I^-$. It is nonvolatile and soluble in water. An aqueous of sodium iodide is an electrolyte.

sodium nitrate $NaNO_3$ (Chile saltpeter) A white soluble deliquescent solid (m.p., 306°C). It decomposes on heating to form sodium nitrite and oxygen; on being strongly heated it forms oxides and peroxides. It is a strong oxidizing agent. Sodium nitrate is used as a fertilizer.

sodium nitrite $NaNO_2$. A yellow hygroscopic crystalline solid; m.p., 271°C; decomposes above 320°C. It is soluble in water. It is used in dyestuffs and to inhibit corrosion.

sodium octadecanoate *See* sodium stearate.

sodium oxides Na_2O (sodium oxide, sodium monoxide) A whiteish gray deliquescent solid that sublimes at 1275°C. It reacts with water, forming sodium hydroxide solution.

Na_2O_2 (sodium peroxide) A white solid that decomposes at 460°C. It reacts with water, forming sodium hydroxide and hydrogen peroxide. Sodium peroxide can absorb carbon dioxide and liberate oxygen and is thus useful in submarines to regenerate the air supply. It is a strong oxidizing agent and is used as a bleaching agent in the textile and paper industries.

NaO_2 (sodium superoxide) A whiteish yellow solid that reacts with water to form a mixture of sodium hydroxide, oxygen, and hydrogen peroxide; a powerful oxidizing agent.

sodium peroxide *See* sodium oxides.

sodium stearate $C_{12}H_{35}COONa$ (sodium octadecanoate) A component of soap.

sodium sulfate Na_2SO_4. A white crystalline solid; m.p., 888°C. It exists in two hydrated forms: the metastable form $Na_2SO_4.7H_2O$, and $Na_2SO_4.10H_2O$ (Glauber's salt), which loses water at 100°C. It is efflorescent, forming the anhydrous salt. All forms are soluble, forming a neutral solution. Sodium sulfate is used in medicine and glass manufacture.

sodium sulfide Na_2S. It has a variable composition; Na_2S_3 and Na_2S_4 are also present, causing a variety of colors such as yellows and reds. It exists in the anhydrous form (m.p., 1180°C) and as $Na_2S.9H_2O$. It is deliquescent and soluble in water, forming an alkaline solution. It is a reducing agent and is corrosive.

sodium sulfite Na_2SO_3. A white, solid, soluble salt that oxidizes readily in air to form sodium sulfate and decomposes on heating to form sodium sulfate and sodium sulfide. It is a reducing agent. It is used in the paper industry to remove chlorine after bleaching.

sodium thiosulfate $Na_2S_2O_3$ ("hypo") A white efflorescent solid that is usually found as the pentahydrate $Na_2S_2O_3.5H_2O$. It is soluble in water, forming a solution that is oxidized in the presence of air. In reactions with dilute acids, sulfur and sulfur dioxide are formed. It is used in photography to fix photographs, and in analytical chemistry.

sol A liquid solution or suspension of a colloid.

solder An alloy used to join metals. It contains different metals, depending on the requirements.

solid A state of matter. In a solid, the particles are not free to move but they can vibrate about fixed positions. Solids can be amorphous or crystalline.

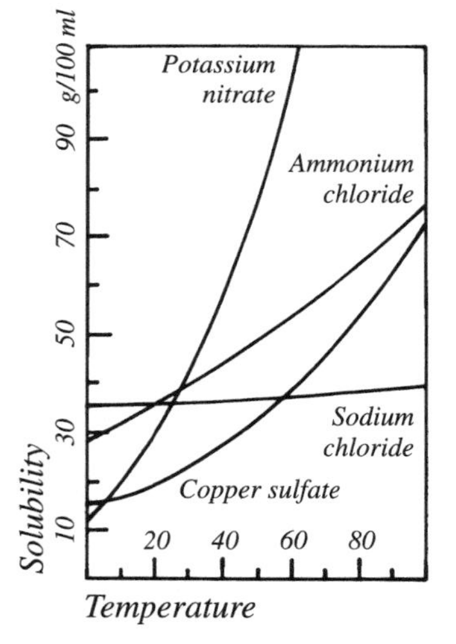

Solubility curve

solubility A measure of the quantity of a solute that will dissolve in a certain amount of solvent to form a saturated solution under certain conditions of temperature and pressure. It is measured in kilograms per meter cubed or moles per kilogram of solvent, etc.

solubility curve A graphic representation of the changing solubility of a solute in a solvent at different temperatures.

solubility of salts All sodium, potassium, and ammonium salts are soluble.

All nitrates are soluble.

All chlorides are soluble except lead chloride, which is soluble in hot water, and silver chloride.

All sulfates are soluble except those of lead, barium, and calcium (which is slightly soluble).

All carbonates are insoluble except those of sodium, potassium, and ammonia. Soluble salts are prepared by crystallization from an aqueous solution. Insoluble salts are prepared by double decomposition.

soluble A relative term that describes a substance that can dissolve in a particular solvent. The extent to which this happens is dependent on temperature.

solute A substance that dissolves in a solvent and thus forms a solution.

solution A uniform mixture of one or more solutes in a solvent. It usually refers to solids dissolved in liquids but can also refer to gases in liquids, gases in solids, etc.

solvation The process of interaction between ions of a solute and the molecules of the solvent. This process is known as hydration when the solvent is water.

Solvay process The production of sodium carbonate from brine (NaCl) and calcium carbonate (limestone) ($CaCO_3$).

As both sodium carbonate and calcium chloride are soluble in water, the process cannot proceed directly. Carbon dioxide is obtained by heating limestone:

$CaCO_3 = CaO + CO_2$.

Ammonia is dissolved in brine and the solution is added to the top of a tower up which carbon dioxide is passed. Ammonium hydrogencarbonate is formed. This reacts to form a precipitate of sodiumhydrogen carbonate, which is sparingly soluble in brine

$NaCl + CO_2 + NH_3 + H_2O = NaHCO_3 + NH_4Cl$. These crystals are collected, purified, and heated to form soda ash (anhydrous sodium carbonate).

$2NaHCO_3 = Na_2CO_3 + CO_2 + H_2O$. The ammonia is recovered from the ammonium chloride produced and is reused.

solvent A substance, usually a liquid, in which a solute dissolves to form a solution.

s orbital A type of orbital whose shape is spherical. One type is possible. It can hold two electrons.

specific gravity *See* relative density.

spectator ions *See* ionic equation.

Sr Symbol for the element strontium.

standardization of solutions Hydrochloric, sulfuric, and nitric acids and the hydroxides of sodium and potassium are either volatile or contain an unknown percentage of water. Standard solutions cannot, therefore, be made up from common acids and alkalis. Their molarities are determined by titration against standard solutions of either sodium carbonate or oxalic acid.

standard solution A solution of known concentration.

starch A polysaccharide (*see* carbohydrate) with the formula $(C_6H_{10}O_5)$. It is composed of many molecules of glucose $(C_6H_{12}O_6)$.

states of matter The three states are solid, liquid, and gas.

state symbols (g) gas; (l) liquid; (s) solid; (c) crystal; and (aq) solution in water.

steam H_2O. Invisible gas formed by water above its boiling point. Clouds of "steam," which are seen before water reaches its boiling point, consist of small droplets of water and are not steam.

steam reforming The conversion of methane, at a temperature of 900°C using a nickel catalyst, into a mixture of carbon monoxide and hydrogen (synthesis gas).

$CH_4 + H_2O = CO + 3H_2$.

stearic acid $C_{17}H_{35}COOH$. A solid saturated fatty acid found in many fats and oils. It is one of the fatty acids used in soap manufacture.

steel An alloy of iron. Other elements are added to form steel of different characters. Mild steel is used for car bodies and household appliances. It rusts easily and is galvanized, enameled, or painted to protect the surface. Stainless steel consists of 74% iron, 18% chromium, 8% nickel. It does not rust and has many uses. Steel used for cutting contains tungsten; steels used at high temperature contain molybdenum.

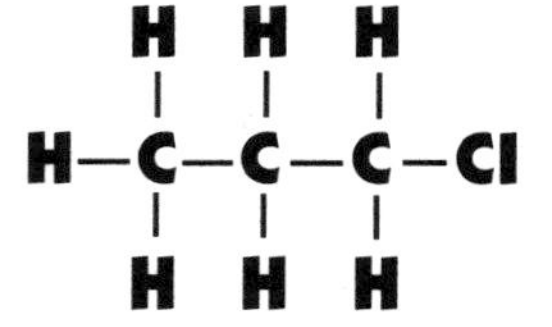

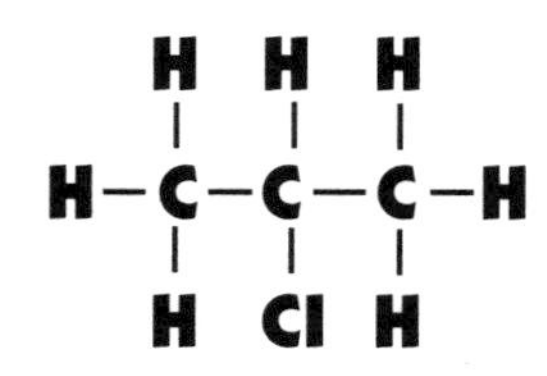

Stereoisomers

steel manufacture Iron is converted to steel by the Bessemer process or in the basic oxygen furnace.

stereoisomerism Isomers of a compound that have the same formula and functional groups and differ only in the arrangements of groups in space are stereoisomers.

stoichiometry Calculations of the proportions in which elements or compounds (molecules) react with each other.

STP Standard temperature and pressure. 0°C (273.16K), pressure 101 325 Pa (760 mmHg).

strengths of acids and bases Acids and bases are considered to be strong if they are fully dissociated into their component ions in solution.

strontium Element symbol, Sr; alkali earth metal, group 2; silvery white metal; Z 38; A(r) 87.62; density (at 20°C), 2.6 g/cm^3; m.p., 769°C; named after Strontian, a Scottish parish; discovered 1808; element used in some alloys and as a vacuum getter. Compounds used to give a red color to fireworks and military flares.

structural formula *See* perspective formula, full structural formula, shortened structural formula.

structural isomer Molecules that are structural isomers have the same molecular formula but have different molecular structures. They may contain different functional groups.

styrene *See* phenylethene.

subatomic particles *See* elementary particles, fundamental particles.

sublimate The solid substance that forms during sublimation – the reversible process by which a substance in a solid state changes directly to a gas. This process can be used to purify a substance.

sublime *See* sublimate.

substitution reaction A typical type of reaction for saturated organic compounds. One or more atoms, or groups of atoms, are replaced by other atoms or groups of atoms.

sucrose $C_{12}H_{22}O_{11}$ (*see* cane sugar) Sucrose is hydrolyzed to form the simple sugars glucose and fructose by the action of dilute acids, or with the enzyme invertase (present in yeast).

$$C_{12}H_{22}O_{11} + H_2O = C_6H_{12}O_6 + C_6H_{12}O_6.$$

sugar (glucose, fructose, sucrose) The general term for members of two groups of sweet-tasting, very soluble carbohydrates – monosaccharides and disaccharides.

sulfates Salts and esters of sulfuric acid (H_2SO_4) containing the ion SO_4^{2-}.

sulfites Salts and esters of sulfurous acid (H_2SO_3) containing the trioxosulfate(IV) ion SO_3^{2-}. Sulfites tend to be reducing agents.

sulfur Element symbol, S; group 6; yellow, nonmetallic, has many allotropic forms; Z 16; A(r) 32.06; density (at 20°C), 2.07 g/cm^3 (rhombic form), 1.96 g/cm^3 (monoclinic form); m.p., 112.8°C (rhombic), 119°C (monoclinic); reactive; name derived from the Latin *sulfur*; known since prehistoric times; used in the manufacture of sulfuric acid and as a plant fungicide.

sulfur dioxide SO_2 (sulfur(IV) oxide) A colorless gas (m.p., –72.7°C; b.p., –10°C) with a pungent odor of burning sulfur. It is very soluble in water, forming a mixture of sulfuric and sulfurous acid. It is a reducing agent. Sulfur dioxide is used in the manufacture of sulfuric acid using the contact process. It is also used as a food preservative and as a bleach.

sulfuric acid H_2SO_4 (tetraoxosulfuric acid, oil of vitriol) It is a strong, oily, colorless, odorless dibasic acid; m.p., 10.36°C; b.p., 338°C. Sulfuric acid is usually used as a 96–98% solution. It is manufactured by the contact process, formerly the lead-chamber process. Concentrated sulfuric acid is a powerful drying agent, and concentrated hot sulfuric

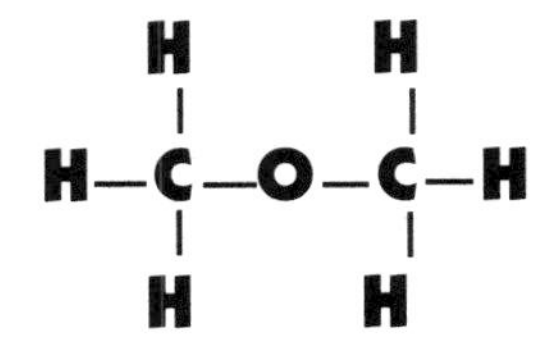

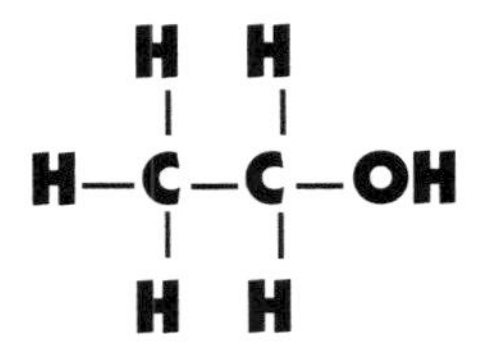

Structural isomers

(methane)

$$CH_3H + XY = CH_3X + HY$$

(ethane)

$$C_2H_5H + XY = C_2H_5X + HY$$

Substitute reactions

acid is a strong oxidizing agent. It forms sulfates and hydrogen sulfates. It is used to make superphosphate for fertilizers.

sulfurous acid H_2SO_3 (sulfuric(IV) acid, trioxosulfuric(IV) acid) It is a weak dibasic acid found only in solution. It forms sulfites and hydrogensulfites and is a reducing agent.

sulfur trioxide SO_3 (sulfur(VI) oxide) A white soluble solid that fumes in moist air. It reacts violently with water to form sulfuric acid. Sulfur trioxide exists in three crystalline forms. It is used in the manufacture of sulfuric acid and oleum.

supercooled vapor A substance that exists as a vapor at a temperature below that at which it should have become liquid.

supercooling The slow cooling of a system, reaching a temperature below that at which a change in phase (liquid to solid or gas to liquid) would normally take place. A supercooled system is in a metastable state.

superheated water Water at a temperature above that of water boiling at one atmosphere.

superheating If a liquid is heated rapidly, its temperature can rise several degrees above the temperature at which it should boil, and it is then said to be superheated. When it does boil, its temperature falls to the boiling point.

superphosphate A mixture of calcium dihydrogen phosphate ($Ca(H_2PO_4)_2$) and calcium sulfate ($CaSO_4$), made by treating calcium(V) phosphate with sulfuric acid and used as a fertilizer as a source of phosphorus.

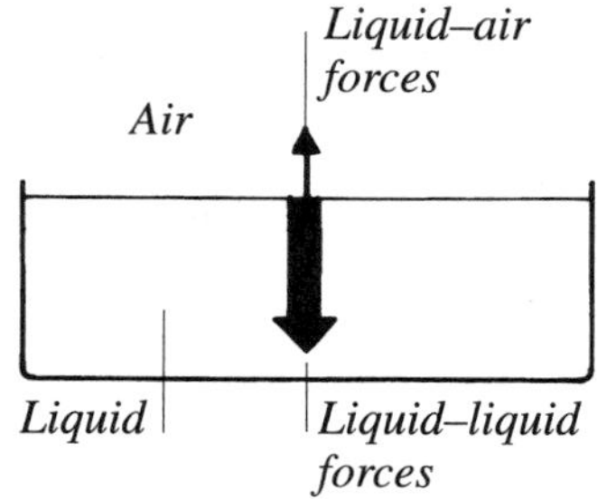

Surface tension

supersaturated solution A solution that contains a higher concentration of solute than does a saturated solution at that temperature. It is usually obtained by cooling a saturated temperature slowly. A supersaturated solution is metastable.

surface active agent A substance (for example a detergent) added to a liquid that can alter its spreading or wetting characteristics by lowering its surface tension.

surface tension Within a liquid, molecules attract each other equally in all directions. At the surface, however, there is no force attracting them outwards, so the molecules are pulled towards the interior of the liquid. For this reason, liquid surfaces tend to become as small as possible.

surfactant Abbreviation for surface active agent.

suspension A type of dispersion. Small solid particles are dispersed in a liquid or gas.

symbols for elements A letter, or group of letters, used to represent the name that has been given to an element.

synthesis The formation of chemical compounds by constructing them directly from their elements or from other simple compounds.

synthesis gas A mixture of carbon monoxide and hydrogen.

synthetic A material that has been prepared artificially rather than being found naturally.

synthetic fibers Fibers that have been prepared artificially (such as rayon from wood pulp, and nylon and polyesters from petroleum derivatives) rather than being produced naturally (cotton, silk, wool).

Ta Symbol for the element tantalum.

talc The white or pale green mineral $Mg_3Si_4O_{10}(OH)_2$. It is soft and greasy and has a hardness of 1 on Mohs' scale. It is often powdered for use in toiletries.

tantalum Element symbol, Ta; transition element; blue-gray metal; Z 73; A(r) 180.95; density (at 20°C), 16.63 g/cm^3; m.p., 2996°C; unreactive; named after *Tantalos*, a mythological Greek character; discovered 1802; used to form resistant alloys used in electronic components and surgical appliances.

tar The heavy thick black liquid that is the residue of the destructive distillation of coal, wood, or petroleum.

tartaric acid HOOC-$(CHOH)_2$(COOH) 2,3-dihydroxybutanedioic acid. A white soluble crystalline organic acid. Its salts are tartrates, and acid salts are hydrogentartarates (*see* cream of tartar). It is found in many plants and fruits, and it is an ingredient of Fehling's solution. Tartaric acid is used in dyeing and in effervescent powders.

tautomerism Tautomerism occurs when a compound exists as two different structural isomers that are in dynamic equilibrium with each other.

Tb Symbol for the element terbium.

Tc Symbol for the element technetium.

Te Symbol for the element tellurium.

technetium Element symbol, Tc; transition element; silvery gray radioactive metal; Z 43; A(r) 98.91; density (at 20°C), 11.5 g/cm^3; m.p., 2172°C; name derived from the Greek *tekhnetos*, "artificial"; discovered 1937.

tellurium Element symbol, Te; group 6; silvery metalloid; Z 52; A(r) 127.6; density (at 20°C), 6.25 g/cm^3; m.p., 450°C; name derived from the Latin *tellus*, "earth"; discovered 1782; used in semiconductors.

temperature A measure of the degree of hotness of a system on a particular scale – Kelvin, degree Celsius, etc.

terbium Element symbol, Tb; rare earth/lanthanide; silvery metal; Z 65; A(r)

158.93; density (at 20°C), 8.23 g/cm^3; m.p., 1356°C; named after Ytterby, Swedish town; discovered 1843; used in electronics.

Terylene *See* Dacron.

test for acid anhydrides If testing a solid, heat it in a test tube. If testing a solution, boil it to dryness and heat the solid residue. If nitrogen dioxide is produced, the substance is a nitrate, peroxide, or dioxide. If carbon dioxide is produced, the substance is a carbonate or bicarbonate. If sulfur dioxide is produced, the substance is a sulfite or bisulfite.

test for chloride ions *See* chlorides.

test to identify carbonate rock Add a few drops of a dilute acid to a rock sample; bubbles of carbon dioxide will be produced if the rock is a carbonate.

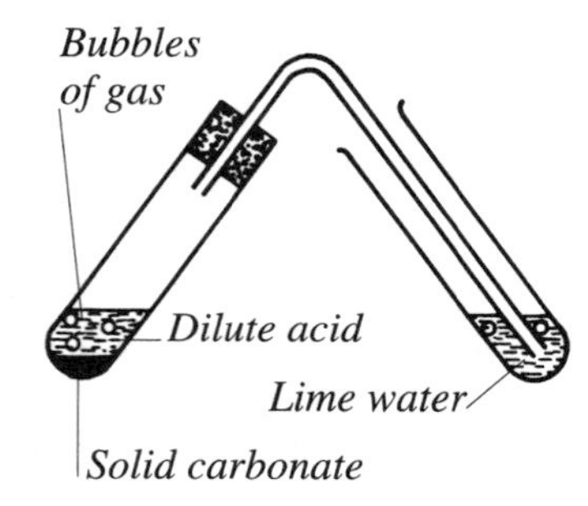

Test to identify carbonate rock (carbon tetrachloride)

tests *See* flame test, blue-ring test for thiosulfate; brown-ring test for nitrates; silver mirror test for aldehydes; tests for reducing sugars; tests for gases; carbohydrates, carbonate, chlorides; bromine test for unsaturated hydrocarbons.

tests for gases Carbon dioxide: pass the gas into lime water. If it is carbon dioxide, the solution turns milky white. Hydrogen: apply a burning taper to the gas. If the gas is hydrogen, it will burn with a pop. Oxygen: apply a glowing taper to the gas. If it is oxygen, the taper will re-light.

tests for reducing sugars (carbohydrates) *See* Fehling's test or (more sensitive) Benedict's test.

tetrachloromethane CCl_4 (carbon tetrachloride) A colorless, poisonous (can penetrate the skin) heavy liquid; m.p., –23°C; b.p., 76.8°C. It is insoluble in water but soluble in all organic solvents. It is formed by the chlorination of methane. It is used as solvent for fats and oils. It was used in dry cleaning and fire extinguishers but is less used now because of its toxicity.

tetraethyl lead $(C_2H_5)_4Pb$. A colorless insoluble viscous liquid. It is an inhibitor that is added to gasoline to prevent premature ignition (*see* antiknock). Its use is being discontinued to minimize lead pollution.

tetrahedral compound A molecule consisting of five atoms, where one atom is at the center of a tetrahedron and the other four atoms are arranged around it at the corners of a tetrahedron, linked by covalent or coordinate bonds. The angle between the bonds is 109° 28'. Carbon compounds form tetrahedral compounds. In methane, a carbon atom is at the center of a tetrahedron and forms covalent bonds to four hydrogen atoms at the corners of a tetrahedron.

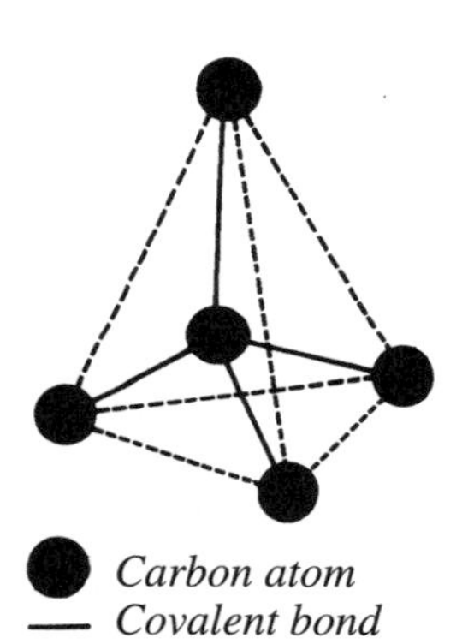

Tetrahedral compound

Th Symbol for the element thorium.

thallium Element symbol, Tl; group 3; grayish metal; Z 81; A(r) 204.37; density (at 20°C), 11.85 g/cm^3; m.p., 303.5°C; name derived from the Greek *thallos*, "green shoot"; discovered 1861.

thermal Relating to heat.

thermal conduction The transfer of heat energy through a substance from a region of high to low temperature. Energy is transferred by vibrations of adjacent molecules. The substance itself does not move.

thermal decomposition The breaking down of a chemical compound by heat into smaller components that do not recombine on cooling.

thermal dissociation The breaking down of a chemical compound by heat into smaller components that recombine to form the original compound on cooling.

thermite process This is a reduction process by which liquid iron can be prepared for welding. It can also be used to prepare liquid chromium and titanium. A mixture of aluminum powder and iron oxide is ignited by a magnesium strip, producing iron and aluminum oxide. The reaction is strongly exothermic and causes the iron formed to melt. It can then be used for welding.

thermodynamic temperature *See* absolute temperature.

thermoplastic A substance (particularly a synthetic plastic) that becomes flexible when heated and hardens on cooling with no change in its properties. Thermoplastic polymers have a molecular chain structure.

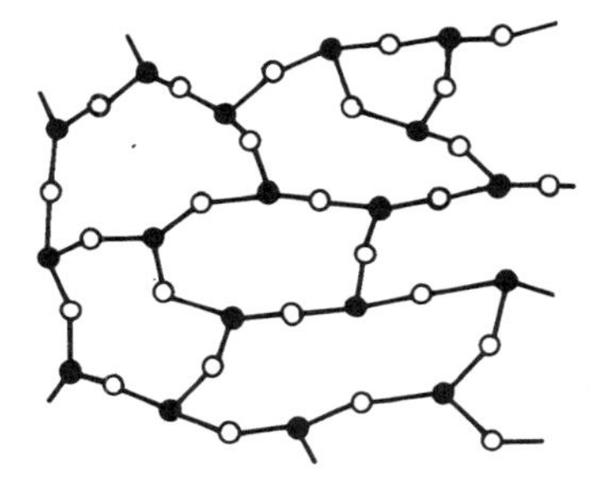

Thermosetting polymer

thermosetting *See* thermosetting polymers, urea/methanal resins.

thermosetting polymers A polymer that has a structure of interlinked chains. Thermosetting polymers cannot be softened by heat but are decomposed by it.

thiosulfate test *See* blue-ring test.

thorium Element symbol, Th; actinide; gray radioactive metal; Z 90; A(r) 232.04; density (at 20°C), 11.73 g/cm^3; m.p., 1750°C; named after Thor, Scandinavian thunder god; discovered 1828; used as fuel in breeder reactors, used as getter. Oxide used for strengthening nickel and as catalyst.

thorium series One of the naturally occurring radioactive series.

thulium Element symbol, Tm; rare earth/lanthanum; soft gray metal; Z 69; A(r) 168.93; density (at 20°C), 9.32 g/cm^3; m.p.,1545°C; name derived from the Latin *Thule*, "Northland"; discovered 1879.

Tl Symbol for the element thallium.

Ti Symbol for the element titanium.

tin Element symbol, Sn; group 4; three allotropes – metallic tin, which is a silvery ductile metal, tetragonal, and gray tin; Z 50; A(r) 118.69; density (at 20°C), 7.3 g/cm^3 (white); m.p., 232°C; Old English name *tin*; symbol name derived from Latin *stannum*; known since prehistoric times; used to coat steel and in alloys (solder and bronze). Compounds used as fungicides.

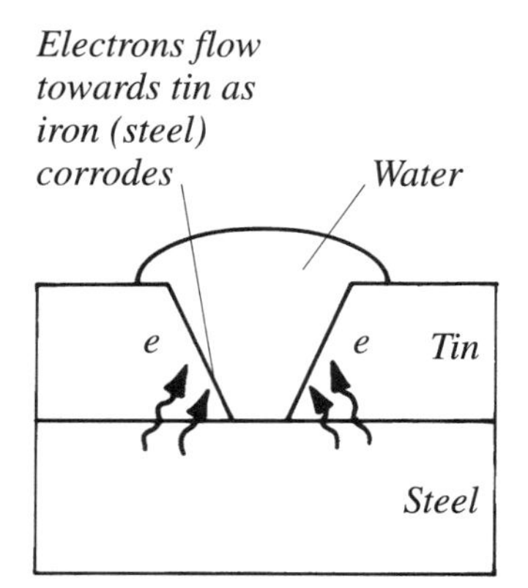

Corrosion of tin cans

tin cans—corrosion of Steel is coated with tin to make cans that hold food. If the surface of the tin is damaged and the steel becomes exposed, the steel rusts very rapidly. This is because iron (steel) is more electropositive than tin, so electrons flow from the iron to the tin, causing the steel to corrode more quickly than if the tin coating were not present. *See* sacrificial anode.

titanium Element symbol, Ti; transition element; white metal; Z 22; A(r) 47.9; density (at 20°C), 4.51 g/cm^3; m.p., 1660°C; named after Titanes, giants in Greek mythology; discovered 1791; is a component of various light, strong alloys used in construction of aircraft, etc. Titanium dioxide is used widely as white pigment in paint.

titration The addition of a solution of known concentration from a burette to a flask containing a known volume of a sample of unknown concentration until the reaction between the two solutions is complete (this point is given by an indicator). The knowledge of the volume of liquid of known concentration added from the burette and the volume of liquid in the flask allows the concentration of the liquid in the flask to be calculated.

Tm Symbol for the element thulium.

Tollen's reagent An aqueous solution containing the complex ion $[Ag(NH_3)_2]^+$. It is prepared by mixing solutions of silver nitrate ($AgNO_3$) and sodium hydroxide (NaOH) to form silver(I) oxide (AgO), which dissolves to form the complex ion $[Ag(NH_3)_2]^+$ when ammonia is added to the solution. This reagent is used in the silver mirror test, which is used to test for the presence of aldehydes.

toluene *See* methylbenzene.

town gas A mixture of coal gas, water gas, and natural gas (a mixture of methane and light gases produced during petroleum refining).

transactinide elements Elements following lawrencium (atomic number 103). This is a transition series; the 6d orbital is being filled. Elements with atomic numbers up to 112 (ununbium, discovered in 1996) have been prepared, and names have been agreed for elements up to 109 (meitnerium, named after Austrian physicist Lise Meitner). Elements 110, 111, and 112 have temporary names. The transactinide elements are unstable and have short half-lives.

transition element Any of the metallic elements with an incomplete inner electron structure.

In the first series of transition elements (scandium to zinc), the elements have two electrons in the s orbital of their fourth shell and d orbitals in their third shell that fill across the row (there can be 10 electrons in d orbitals).

In the second series (yttrium to cadmium), the elements have two electrons in the s orbital of their fifth shell and d orbitals of their fourth shell, which fill across the row.

The third series begins with lanthanum and ends with mercury (it includes the lanthanides). As the atoms increase in size, their structure becomes more complex as the number of types of orbital in each shell becomes greater.

The elements from actinium onwards (the actinides and transactinides) can also be considered as transition elements.

Transition elements have widely differing chemical properties. They can each have several oxidation numbers (oxidation states), and they form colored compounds.

transition metals *See* transition elements.

transition point *See* transition temperature.

transition temperature The temperature at which one allotrope changes to another (one is stable below the transition temperature, the other is stable above it). It can also be the temperature at which a substance changes phase.

transuranic element Any of the elements with higher atomic numbers than uranium. They are all radioactive and are produced artificially.

tribasic acid An acid that has three replaceable hydrogen atoms. A tribasic acid can form three series of salts. *See* basicity of acids.

trichloroethane $CHCl_3$ (chloroform). A heavy, colorless volatile liquid that is toxic, non-flammable, and insoluble in water. It is a substituted alkane; m.p., –63.5°C, b.p., 61°C. It can be made by chlorination of methane, followed by separation of the products. It has been widely used as an anesthetic but can cause liver damage; it has been superseded by other halogenated hydrocarbons.

triple bond *See* bond, multiple bonds.

triple point The conditions of temperature and pressure at which the three phases of a substance – solid, liquid, and gas – are in equilibrium.

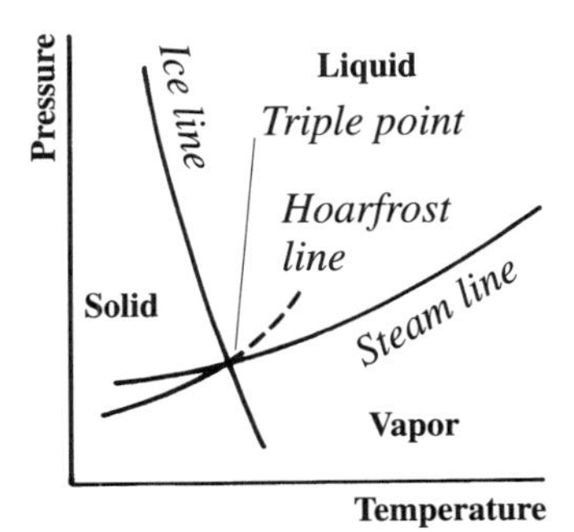

Triple point

tritium An isotope of hydrogen. Its nucleus contains one proton and two neutrons and thus has a relative atomic mass of three.

trivalent Having a valency of three.

tungsten Element symbol, W; transition element; white or gray metal; Z 74; A(r) 183.85; density (at 20°C), 19.3 g/cm^3; m.p., 3410°C; name derived from the Swedish *tung sten*, “heavy stone”; symbol name derived from German *Wolfram*; discovered 1781; used in steel alloys used to make cutting tools, in filaments of lamps and heaters.

U Symbol for the element uranium.

universal indicator A mixture of substances that can be used as an acid-base indicator over a wide range of pH values. Its color changes from red (pH values 1–4), orange (pH 5), yellow (pH 6), green (pH 7), blue (pH 8), indigo (pH 9), violet (pH 10–14).

unleaded gasoline Gasoline that does not contain tetraethyl lead to prevent knocking (*see* antiknock). In its place, it contains 15% methyl tertiary butyl ether and 5% methanol.

unsaturated compounds Chemical compounds that contain one or more double or triple bonds in their structure.

Unsaturated hydrocarbons contain double or triple bonds between carbon atoms in their structure. Ethene and ethyne are examples of unsaturated hydrocarbons. Unsaturated hydrocarbons take part in addition reactions when a double bond is converted to a single bond or a triple bond is converted to a double or a single bond.

unsaturated hydrocarbons—test *See* bromine test.

unsaturated solution A solution in which the solvent is able to absorb more solute at a particular temperature.

ununbium Element symbol, Uub; transition element; Z112; A(r) 277; temporary name; discovered 1996.

ununnilium Element symbol, Uun; transition element; Z 110; A(r) 269; temporary name; discovered 1994.

unununium Element symbol, Uuu; transition element; Z 111; A(r) 272; temporary name; discovered 1994.

uranium Element symbol, U; actinide; white radioactive metal; Z 92; A(r) 238.03; density (at 20°C), 19.05 g/cm^3; m.p., 1132°C; three isotopes, 234, 235, and 238; named after the planet Uranus; discovered 1789; isotope-235 used in nuclear reactors and nuclear weapons.

uranium series One of the naturally occurring radioactive series.

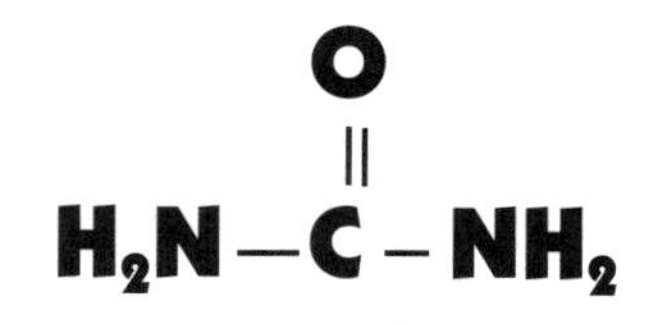

Urea (carbamide)

urea (carbamide) NH_2–CO–NH_2. White crystalline soluble solid; m.p., 133°C.

urea/methanal resin Thermosetting polymer made by condensation polymerization between urea and methanal. It is used to form melamine and to adhesives.

Uub Symbol for the element ununbium.

Uun Symbol for the element ununnilium.

Uuu Symbol for the element unununium.

V Symbol for the element vanadium.

valency It is the number of electrons an atom needs to form a compound or radical and is related to its electronic structure. Elements tend to lose, gain or share electrons in order to complete their outer electron shell. An ion's valency is equal to its charge. Valency is also seen as the usual number of bonds an atom forms when combining to form a compound.

vanadium Element symbol, V; transition element; silvery white or gray metal; Z 23; A(r) 50.94; density (at 20°C), 6.09 g/cm^3; m.p., 1890°C; named after Vanadis, Scandinavian goddess; discovered 1801; used as steel additive and in catalysts.

van der Waals forces Weak intermolecular or interatomic forces between neutral molecules or atoms. They are much weaker than chemical bonds.

vapor Gas that is below the temperature at which it can be liquefied by pressure (the critical temperature).

vapor density The vapor density of a vapor (gas) is the ratio of the mass of a certain volume of the gas to the mass of an equal volume of hydrogen that is at the same temperature and pressure. If the density of hydrogen is taken as one, the vapor density of a gas is half its relative molecular mass.

vaporization The process of change of state of a solid or liquid to a vapor.

vapor pressure The pressure exerted by a vapor given off by a liquid or solid. If this vapor is in equilibrium with the liquid or solid, the vapor pressure is defined as the saturated vapor pressure.

variable A condition, such as temperature, concentration, or pressure, that can be changed in a chemical reaction.

volatile A substance that readily turns into a vapor.

volumetric analysis A method of quantitative analysis such as titration, where the volume of a solution reacting with a measured amount of another solution is determined.

vulcanization The process of making rubber harder and more elastic. Rubber is heated with sulfur (about 5%) at about 150°C. This causes the rubber molecules to become interlinked into a three-dimensional network with the sulfur molecules forming cross-links between adjacent rubber molecules.

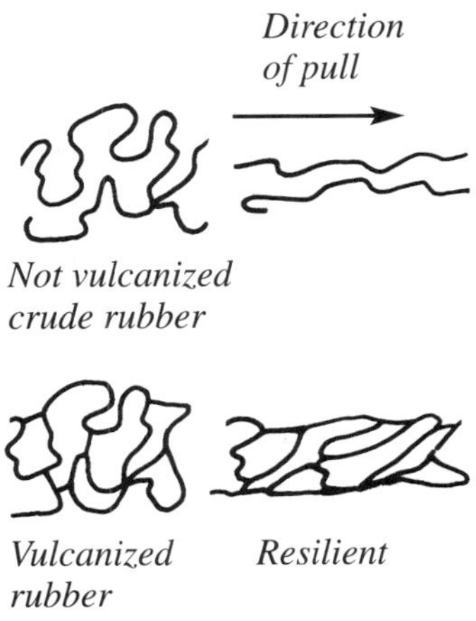

Direction of pull

Not vulcanized crude rubber

Vulcanized rubber

Resilient

Vulcanization

W Symbol for the element tungsten.

washing soda *See* sodium carbonate.

water H_2O. A colorless, odorless, tasteless liquid; m.p., 0°C, b.p., 100°C. An oxide of hydrogen, it is a covalent compound. The three atoms do not lie in a straight line H-O-H; the position of the two hydrogen atoms is affected by the two lone pairs of electrons in the outer shell of the oxygen atom and, therefore, form an angle of about 105° with the oxygen atom. Water molecules are polar, and there is hydrogen bonding between the molecules. Because of its polar nature, water is an excellent solvent for ionic substances.

water—electrolysis of If slightly acidified, water can be electrolyzed between carbon or platinum electrodes. Gases are evolved above the electrodes. Hydrogen is collected at the cathode (the negative electrode; positive hydrogen ions gain an electron each). Oxygen is collected at the anode (negatively charged hydroxide ions lose an electron, forming oxygen and water).

water gas Water gas is a mixture of approximately equal volumes of hydrogen and carbon monoxide. Formed by passing steam (rather than air, as in producer gas manufacture) through white hot coke, it has a high calorific value.

$C + H_2O(g) = CO + H_2 -$ heat.

$CO + H_2O - CO_2 + H_2$.

Water gas is a cheap source of commercial hydrogen. Methanol is manufactured from water gas and hydrogen. Water gas is added to coal gas to increase its calorific value.

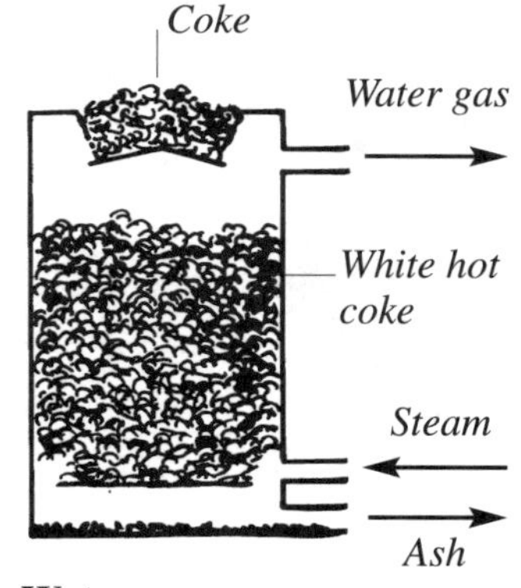

Water gas

water ionization A water molecule can split up to form a positively charged hydrogen ion and a negatively charged hydroxide ion.

water of crystallization The exact number of water molecules that are chemically bonded to a molecule of a salt within a hydrated crystalline compound.

water pH Pure water is a neutral liquid (pH = 7). It consists almost entirely of covalent molecules.

water—test for As no two substances have the same freezing points and boiling points, water can be identified as freezing at 0°C and boiling at 100°C.

weak acids and bases Acids and bases are considered to be weak if they do not dissociate into their component ions in solution.

white lead *See* lead(II) carbonate hydroxide.

white vitriol Hydrated zinc sulfate ($ZnSO_4.7H_2O$) or zinc sulfate heptahydrate.

word equation A summary in words of the reactants and products taking part in a chemical reaction.

Xe Symbol for the element xenon.

xenon Element symbol, Xe; noble gas, group 8; colorless gas; Z 54; A(r) 131.3; density (at 20°C), 5.896 g/l at STP; m.p., –111.9°C; name derived from the Greek *xenos*, "strange"; discovered 1898; used in fluorescent lamps and bubble chambers.

Y Symbol for the element yttrium.

Yb Symbol for the element ytterbium.

yield of a reaction Many chemical reactions produce less product than might be predicted from the equation of the reaction. The yield of a reaction is the amount of product produced in a reaction expressed as a percentage of the theoretical yield.

ytterbium Element symbol, Yb; rare earth/lanthanide; silvery metal; Z 70; A(r) 173.04; density (at 20°C), 6.97 g/cm^3; m.p., 819°C; named after Ytterby, a Swedish town; discovered 1907; used in some special steels.

yttrium Element symbol, Y; transition element; silvery gray metal; Z 39; A(r) 88.91; density (at 20°C), 4.47 g/cm^3; m.p., 1522°C; named after Ytterby, a Swedish town; discovered 1828; alloyed with cobalt to make superconducting alloys and strong permanent magnets. Oxide used in color televisions.

Z Symbol for atomic number.

zeolite A naturally occurring mineral form of sodium aluminum silicate. It has ion-exchange properties.

zeolites Either naturally occurring minerals or synthetic substances that have ion-exchange properties. They can be hydrated silicates of aluminum, sodium, potassium, or calcium. Alternatively, they can be constructed to form molecular sieves because their structures have open pores to trap certain molecules from a mixture passing through them; the molecules are then released by heating the zeolite.

zinc Element symbol, Zn; transition element; hard, brittle bluish white metal; Z 30; A(r) 65.37; density (at 20°C), 7.1 g/cm^3; m.p., 419.6°C; chemically reactive; releases hydrogen from dilute acids; name derived from the German *Zink*; discovered before 1300; used as sacrificial anodes; used in alloys (brass, solder) and to coat steel surfaces (galvanizing).

zinc blende A naturally occurring mineral form of zinc sulfide (ZnS) from which zinc is extracted.

zinc carbonate $ZnCO_3$. A white insoluble crystalline compound that is found in the mineral calamine. Zinc carbonate is used to make zinc ointments.

zinc chloride $ZnCl_2$. A white crystalline soluble compound; m.p., 290°C; b.p., 732°C (it sublimates easily). The anhydrous salt is deliquescent and is used as a dehydrating agent. It is also used as a flux in soldering and as a timber preservative.

zinc oxide ZnO. An insoluble powder that is white when cold and yellow when hot (on heating, it loses a small amount of oxygen, which it reabsorbs on cooling); m.p., 1975°C. It is found naturally as the ore zincite. It is an amphoteric base. Zinc oxide is used as the white pigment zinc white in the glass and ceramics industries and (it is a mild antiseptic) in antiseptic ointments.

zinc sulfate $ZnSO_4$. A white crystalline water-soluble compound, formerly known as white vitriol in its heptahydrate ($ZnSO_4.7H_2O$) form. It is used as a mordant and to check bleeding (as a styptic).

zinc sulfide ZnS. A yellow-white compound that is phosphorescent when impure. It sublimes at 1180°C. It is used as a pigment and as a phosphor.

zirconium Element symbol, Zr; transition element; gray-white metal; Z 40; A(r) 91.22; density (at 20°C), 6.5 g/cm^3; m.p., 1852°C; low neutron absorption; name derived from the German *Zirkon*; discovered 1789; alloys used in construction of reactors.

Zn Symbol for the element zinc.

Zr Symbol for the element zirconium.

○ S^{2-}
● ZN^{2+}

Zinc sulphide

SECTION TWO BIOGRAPHIES

Abel, Sir Frederick Augustus (1827–1902) English chemist who took up an appointment with the War Department and specialized in the chemistry of explosives. He found a way to make guncotton stable and safe and, in 1889, introduced the new explosive cordite, a mixture of nitroglycerine and guncotton stabilized with camphor. This was used extensively in World War I.

Abelson, Philip Hauge (b. 1913) American physical chemist who developed a massive gas diffusion apparatus for the separation of the fissionable uranium-235 isotope from the natural mixture, which was almost all uranium-238. This was an early stage in the production of the first atomic bomb. Abelson also assisted in the creation of the manufactured element neptunium, the first element heavier than uranium. Later he worked with Miller to try to show in the laboratory how life might have originated on Earth.

Adams, Roger (1889–1971) American chemist who developed a simple way of catalyzing the hydrogenation of unsaturated organic materials, such as vegetable oils (to make margarine and other butter substitutes). He established the molecular structure of various medically active natural substances and isolated tetrahydrocannabinol, the active ingredient in marijuana.

Alder, Kurt (1902–58) German chemist who, in 1928, working with Diels, discovered a relatively easy way to produce a ring (cyclic) compound, starting with a compound containing two double bonds separated by a single bond. This is the Diels-Alder reaction, which became important in organic synthesis and earned Alder a share in the 1950 Nobel Prize in chemistry.

Anaxagoras (c.500–428 BCE) Greek philosopher who was persecuted for believing that the Sun was an incandescent rock. He taught that matter was composed of innumerable tiny particles containing determining qualities.

Anaximenes (fl. c.500 BCE) Greek philosopher who believed that the origin of all matter was air and that this could be condensed to make various forms of solid matter or liquids. He believed that the Earth was flat.

Anfinsen, Christian B. (b. 1916) American biochemist who assisted in the sequencing of the 128-amino acid enzyme ribonuclease, an achievement for which he shared the 1972 Nobel Prize in chemistry. Anfinsen went on to study the three-dimensional (secondary and tertiary) structure of this important enzyme.

Christian B. Anfinsen

Archimedes (c.287–212 BCE) Greek mathematician and technologist who found formulas for the volume of a wide range of regular solids and for the area of a range of plane figures. His methods were similar to those of the calculus. He originated the science of hydrostatics and discovered that a floating body displaced its own weight of water. He invented the Archimedian screw for raising water and a number of large military weapons. He was the leading figure in rigorous scientific and mathematical thought of the ancient world.

Archimedes

Aristotle (384–322 BCE) Greek philosopher who, through his extensive writings, became historically the most influential figure of the ancient world. He covered every field of contemporary knowledge, but in science wrote on physics, biology, medicine, zoology, taxonomy, and psychology. Much of what he wrote about fundamental science was pure imagination and was wrong. Unfortunately, he was accepted as an almost infallible authority, so for almost 2000 years, scientific thought was misdirected and real progress hindered.

Arrhenius, Svante (1859–1927) Swedish physical chemist who, in 1884, was the first to propose that acids, bases, and salts in solution dissociated into ions. His theory of electrolytic dissociation was well before its time and was not scientifically confirmed until the theory of atomic structure was more fully developed. He also worked on reaction rates, and was the first to recognize the "greenhouse effect" on climate. He was awarded the Nobel Prize in chemistry in 1903.

Aston, Francis (1877–1945) English atomic physicist who worked with J. J. Thomson at the Cavendish Laboratory, Cambridge. His principal field of study was in elements of equal atomic number but different atomic weight (isotopes). He also invented the mass spectrograph. He was awarded the Nobel Prize in chemistry in 1922.

Avogadro, Amedeo (1776–1856) Italian scientist and professor of physics at Turin (1834–50) who in 1811 formulated the hypothesis known as Avogadro's law: equal volumes of gases contain equal numbers of molecules, when at the same temperature and pressure.

Axelrod, Julius (b. 1912) U.S. neuropharmacologist who was a member of the research team that discovered the neurotransmitter norepinephrine (noradrenaline). Axelrod was refused entry to several medical schools and decided to study pharmacology. In 1970, he and two of his colleagues were awarded the Nobel Prize in physiology or medicine.

Francis Bacon

Bacon, Francis (1561–1626) English philosopher and essayist whose book *The Advancement of Learning* (1605) drew serious attention for the first time to the fact that the real source of scientific knowledge was not the authority of pundits such as Aristotle but was observation, experimentation, direct experience, and careful induction. This was the start of the scientific method that was to prove so fruitful.

Bacon, Roger (c.1214–92) English philosopher who tried to compile an encyclopedia containing all the knowledge of his day. The attempt failed but contained much valuable mathematical information, showed a knowledge of gunpowder, and included some remarkable speculations about mechanical transport, heavier-than-air flying machines, and the possibility of circling the globe.

Baekeland, Leo Hendrik (1863–1944) Belgian-born American industrial chemist who became a millionaire when he sold his Velox photographic paper company to Kodak. He then studied chemistry and investigated phenol-formaldehyde resins and produced a hard material that could be cast and machined and was a good electrical insulator. He called it Bakelite.

Friedrich Adolf von Baeyer

von Baeyer, Friedrich Adolf (1835–1917) German chemist who devoted his life to the analysis and synthesis of organic molecules and published more than 300 important papers. He is especially noted for his studies of uric acid and organic dyes. His synthesis of indigo was of great commercial importance, and for this achievement he was awarded the Nobel Prize in chemistry in 1905.

Balard, Antoine-Jérôme (1802–76) French chemist who discovered that iodine produces a blue color in the presence of starch, a finding that produced a sensitive chemical test for iodine that is still in use today.

Antoine-Jérôme Balard

Balmer, Johann Jakob (1825–98) Swiss mathematician who, in 1885, empirically derived a simple formula for the wavelengths of the spectral lines of hydrogen. This is called the Balmer series, and Balmer could not explain it, but its quantitative element was important in developing atomic theory. Later, better models of the atom accounted for the series.

Baltimore, David (b. 1938) American biochemist who shared the 1975 Nobel Prize in medicine with Howard Temin and Renato Dulbecco for discovering an enzyme, called reverse transcriptase, that could make DNA from RNA. This viral enzyme, present in retroviruses such as HIV, enables these viruses to insert their genome into the DNA of the host cell. Its discovery showed that Crick's fundamental genetic "dogma" – that the sequence is always from DNA to RNA to protein – was wrong.

David Baltimore

Bamberger, Eugen (1857–1932) German chemist who first proposed the term "alicyclic" for unsaturated ring organic compounds. He worked on the synthesis of nitroso compounds and quinols and investigated the structure of naphthalene.

Barger, George (1878–1939) Dutch-born British organic chemist who isolated ergotoxine from ergot and proceeded to study related amines with physiological properties. This eventually had two important effects: it drew attention to the role of neurotransmitters in the function of the nervous system, and it led to the development of a range of valuable drugs.

Bartlett, Neil (b. 1932) English chemist who was the first to find a compound of a noble gas (xenon), one of a class of elements that was previously believed incapable of forming compounds.

Bartlett, Paul D. (b. 1907) American chemist who worked on the mechanisms of organic reactions such as the actions of free radicals and the way polymer molecules were formed from simple units (monomers).

Barton, Sir Derek H. R. (1918–98) English chemist who became noted when he wrote an influential paper on the relationship between the three-dimensional shape of a molecule, resulting from rotation of part of it around a single bond, and its chemical reactivity. This significant and fundamental advance won him the Nobel Prize in chemistry in 1969, which he shared with Hassel.

Beadle, George Wells (1903–89) U.S. biochemist who showed that particular genes code for particular enzymes. The technique used was to cause gene mutations that affected particular biochemical processes, thereby showing the immediate link between the gene and the enzyme. This was a major advance in genetics.

Becher, Johann Joachim (1635–82) German chemist who studied minerals. His *Physica subterranea* (1669) was the first attempt to bring physics and chemistry into close relation.

Johann Joachim Becher

Becquerel, Antoine Henri (1852–1908) French physicist who was the first to discover radioactivity. In 1896 he noticed that a uranium salt laid on a totally enclosed photographic plate caused the plate to be exposed. Becquerel concluded that the salt was emitting rays similar to the X rays that had been discovered byWilhelm Roentgen the year before. He then studied and described the properties of the natural radioactivity of uranium. For this work, he shared the Nobel Prize in physics with Marie Curie in 1903.

Bednorz, Johannes Georg (b. 1950) German physicist who worked with Alex Muller on the problem of raising the temperature at which superconductivity occurred. Applications of superconductivity were seriously limited because of the energy required to maintain temperatures close to absolute zero. Bednorz and Muller came up with a mixture of lanthanum, barium, and copper oxide that would superconduct at 35K. This was a substantially higher temperature than with any previous material, and it won the two men the 1987 Nobel Prize in physics.

Berg, Paul (b. 1926) American biochemist and molecular biologist who was one of the principal founders of genetic engineering. Berg developed techniques using specific DNA-cleaving enzymes capable of cutting out genes from the DNA of one mammalian species and inserting them into the DNA of another. For this work, he shared the 1980 Nobel Prize in chemistry with Walter Gilbert and Sanger. Berg also drew up strict rules to govern safe conduct in genetic engineering.

Bergius, Friedrich (1884–1949) German chemist who demonstrated the way in which high pressures and temperatures converted wood into coal. For this work, he was awarded the 1931 Nobel Prize in chemistry.

Bernard, Claude (1813–78) French physiologist, often described as the father of modern physiology. After becoming a doctor, he devoted his life to research in physiology, eventually establishing it as a formal discipline in its own right. Among his many discoveries were that complex food carbohydrates were broken down to simple sugars before absorption; that bile was necessary for the absorption of fats; that the body re-synthesizes the carbohydrate glycogen; and that the bore of arteries was controlled by nerve action. His later research included work on oxygen in the blood and the opium alkaloids.

Claude Bernard

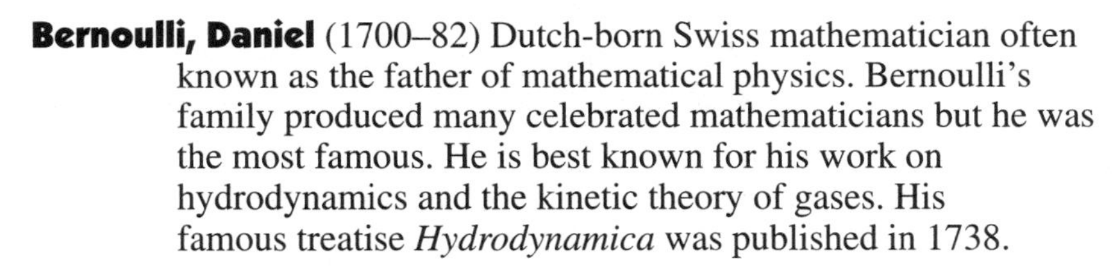

Bernoulli, Daniel (1700–82) Dutch-born Swiss mathematician often known as the father of mathematical physics. Bernoulli's family produced many celebrated mathematicians but he was the most famous. He is best known for his work on hydrodynamics and the kinetic theory of gases. His famous treatise *Hydrodynamica* was published in 1738.

Berthollet, Claude Louis (1748–1822) French physician and inorganic chemist whose ideas on the formation of chemical compounds, although mistaken, led other workers to come nearer to the truth. He made important advances in dye-making, bleaching with chlorine, and steel making. His two major published works were *Researches into the Laws of Chemical Affinity* (1801) and *Essay on Chemical Statics* (1803).

Claude Louis Berthollet

Berzelius, Jons Jacob (1779–1848) Swedish chemist who was able to work out the atomic weights of more than 45 elements, several of which he discovered, including, cerium, selenium, thorium, and vanadium. He proposed the theories of isomerism and catalysis and was also notable for inventing the present-day symbols for chemical elements and compounds based on abbreviations of the Latin names of the elements.

Jons Jacob Berzelius

Sir Henry Bessemer

Bessemer, Sir Henry (1813–98) English chemist, inventor, and engineer. In 1855, in response to the need for guns for the Crimean War, he patented the process by which molten pig-iron can be turned directly into steel by blowing air through it in a Bessemer converter. This development of a cheap way to produce steel had enormous economic importance and won him a knighthood in 1879. Bessemer furnaces were enthusiastically exploited in the United States by Andrew Carnegie, where they made a fortune for him.

Bevan, Edward John (1856–1921) English industrial chemist who developed the viscous process of rayon (processed cellulose) manufacture. This proved commercially highly profitable until better plastics such as nylon were developed.

Joseph Black

Black, Sir James Whyte (b. 1924) Scottish physician and physiologist who studied cell receptors for hormones and drugs and developed two new classes of drugs – the beta blockers (for example, Sectral) and the stomach histamine (H-2) receptor blockers (for example, Zantac). Black, who is remarkable for initiating two entirely new classes of major drugs, was awarded the 1988 Nobel Prize in physiology or medicine.

Black, Joseph (1728–99) Scottish chemist who between 1756 and 1761 evolved the theory of "latent heat" on which his scientific fame chiefly rests. He also showed that the causticity of lime and alkalis is due to the absence of the "fixed air" (carbon dioxide) that is present in limestone and carbonates of alkalis.

de Boisbaudran, Paul-Émile Lecoq (1838–1912) French chemist and wine merchant whose main scientific work was in spectroscopic analysis. He found a new element, gallium, and helped in the discovery of five of the fifteen rare earth elements (lanthanoids).

Boltwood, Bertram Borden (1870–1927) American nuclear chemist who studied the radioactive breakdown of elements, and first discovered how to apply the ratios of lead to uranium in geological specimens in order to calculate their age. This, and other associated methods, was to bring new standards of accuracy into geology and paleontology. He discovered the radioactive element ionium.

Bonner, James Frederick (b. 1910) American molecular biologist who showed that the presence of a protein, histone, shut down gene activity so that only those genes required in a particular situation were operative. Bonner also worked on the artificial synthesis of RNA.

Carl Bosch

Bosch, Carl (1874–1940) German chemist who invented the first commercially successful method of synthesizing ammonia on an industrial scale. This was of great importance for agriculture and earned him a share of the 1931 Nobel Prize in chemistry. He also worked on the synthesis of methanol and rubber.

Daniel Bovet

Bovet, Daniel (1907–92) Swiss pharmacologist and physiologist who isolated the sulfanilamide part of the red dye prontosil that was being used to treat bacterial infections in mice. This work led to the development of the sulfa drugs, which, until the development of penicillin and other antibiotics, were the most important class of medications.

Boyer, Herbert W. (b. 1936) American biochemist and genetic engineering pioneer who showed that, by using an enzyme called an endonuclease, a DNA ring (a plasmid) from a bacterium could be inserted into the DNA of another bacterium or that of a toad. He was then able to clone the hybrid DNA by allowing bacteria containing it to reproduce. Boyer's achievement led to the commercial production of insulin and other valuable proteins.

Boyer, Paul D. (b. 1918) American chemist whose advances in the understanding of the mechanism by which the enzyme ATP synthase (ATPase) catalyzes the synthesis of ATP from ADP and phosphate earned him a share of the 1997 Nobel Prize in chemistry. Adenosine triphosphate (ATP) is a nucleotide of fundamental importance as the carrier of chemical energy in all living organisms.

Robert Boyle

Boyle, Robert (1627–91) Irish physicist, chemist, and cofounder of the Royal Society in London whose book *Sceptical Chymist* (1661) defines the chemical element as the practical limit of chemical analysis. The celebrated Boyle's law (1662) states that, the temperature being kept constant, the pressure and volume of a gas are inversely proportional.

Brand, Hennig (d. 1692) German alchemist who, in 1669 while searching for the philosopher's stone, isolated from urine a solid white substance that glowed in the dark. He called it phosphorus. Brand, whose date of birth is unknown, is the first scientist known to have discovered an element.

Brandt, Georg (1694–1768) Swedish chemist and assay master of the Swedish mint who discovered cobalt, studied arsenic and established its properties and compounds, and worked to expose fraud in alchemical practice.

Brown, Herbert C. (b. 1912) American chemist whose work on the use of boron compounds in the synthesis of organic molecules won him the 1979 Nobel Prize in chemistry.

Buchner, Eduard (1860–1917) German chemist who researched fermentation and showed that Pasteur was wrong in insisting that alcoholic fermentation required the exclusion of oxygen. For this finding he was awarded the 1907 Nobel Prize in chemistry.

Robert Wilhelm Bunsen

Bunsen, Robert Wilhelm (1811–99) German experimental chemist and inventor. He developed the gas burner that bears his name and the ice calorimeter. Working with the German physicist Gustav Kirchhoff, he developed the important analytical technique of chemical spectroscopy. Bunsen also discovered the elements cesium and rubidium.

Butenandt, Adolf F. J. (1903–95) German biochemist who isolated the male sex hormone androsterone and, in 1931, a few milligrams of progesterone from the corpus luteum of the ovaries of female pigs. His methods must have been relatively inefficient, as no fewer than 50,000 pigs were required. For his work on hormones he shared the 1939 Nobel Prize in chemistry.

Calvin, Melvin (1911–97) American biochemist who made notable advances in the understanding of photosynthesis – the processes by which sugars and complex carbohydrates such as starches are synthesized by plants from atmospheric carbon dioxide. Calvin used radioactive carbon tracers to follow the movement of carbon through the complex reactions. He also worked on theories of the chemical origin of life and on attempts to utilize carbon dioxide artificially. He was awarded the Nobel Prize in chemistry in 1961 for his work on photosynthesis.

de Candolle, Augustin Pyrame (1778–1841) Swiss botanist and chemist who introduced the term "taxonomy" for the classification of plants by their morphology, rather than physiology, as set out in his *Elementary Theory of Botany* (1813). His new edition of *Flore française* appeared in 1805. He accurately described the relationship between plants and soils, a factor that affects geographic plant distribution. He is remembered in the specific names of more than 300 plants, two genera, and one family.

Augustin Pyrame de Candolle

Cannizzaro, Stanislao (1826–1910) Italian chemist who showed that inorganic and organic chemistry were not basically different. He made the important distinction between atomic weights and molecular weights, and produced a table of weights with hydrogen as the unit. He established the use of atomic weights in chemical formulas and calculations.

Cardano, Geronimo (1501–76) Italian mathematician and physician who produced the first printed work on algebra, the *Ars magna* (1545), and described the general method for solving cubic equations. He also published books on medicine, arithmetic, and philosophy and an encyclopedia of inventions and experiments called *De subtilitate rerum* (1550).

Carnap, Rudolf (1891–1970) German philosopher of science who has had considerable influence on scientists. He taught that statements were meaningful only if they can be related to sensory experience and have logical consequences that are verifiable by observation or experience. This idea led to the philosophical school of logical positivism and to the dismissal, by some people, of most or all of the propositions of metaphysics and religion.

Carothers, Wallace Hume (1896–1937) American industrial chemist working for the Du Pont Company at Wilmington, Delaware, who invented nylon after producing the first successful synthetic rubber, neoprene.

Castner, Hamilton Young (1858–98) American chemist who discovered a cheap way to produce metallic sodium from caustic soda so that it could be used to reduce aluminum chloride to the metal. Unfortunately for him, the electrolytic process for aluminum production was, just then, invented. Castner went through a difficult period until the uses of sodium for other purposes increased to the point where he could not produce enough to meet the demand. So he invented an even better process for producing sodium from salt water by electrolysis. He formed a company that became highly successful after his death.

Henry Cavendish

Cavendish, Henry (1731–1810) English chemist and natural philosopher. In 1760, he discovered the extreme levity of inflammable air, and later, at the same time as James Watt, ascertained that water is the result of the union of two gases. He used the gravitational attraction between bodies of known weight to estimate the weight of the Earth.

Caventou, Jean Bienaimé (1795–1877) French chemist and toxicologist who specialized in alkaloids. Working with Pelletier, he isolated quinine, strychnine, brucine, cinchonine, veratrine, and colchicine, some of which became widely used as drugs.

Thomas R. Cech

Cech, Thomas R. (b. 1947) American biochemist who showed the remarkable fact that a length of protein-free RNA could act as an enzyme for the cleaving and splicing of other RNA. This fact could explain much about the early evolution of organisms. Cech shared the 1989 Nobel Prize in chemistry with Sidney Altman.

Sir Ernst Boris Chain

Chain, Sir Ernst Boris (1906–79) German-born British biochemist of Russian extraction who isolated and purified penicillin and turned Fleming's discovery of the antibiotic into one of the greatest successes in the history of medicine. Chain, Florey, and Fleming shared the 1945 Nobel Prize in physiology or medicine.

Chance, Britton (b. 1913) American biophysicist who proved that enzymes work by attaching themselves to the substance on which they act (the substrate). He achieved this by a spectroscopic technique using the enzyme peroxidase, which contains iron and absorbs certain light wavelengths strongly. Chance also helped to work out the way cells get their energy from sugar by observing that concentrations of adenosine diphosphate (ADP) were related to the oxidation-reduction states of the proteins in the respiratory chain.

Chargaff, Erwin (b. 1905) Austrian-born American biochemist who, in the mid-1940s, speculated that if DNA was the vehicle of inheritance, its molecule must vary greatly. Using the methods available at the time, however, he found that its composition was constant within a species but that it differed widely between species. In 1950, he established that the number of purine bases (adenine and guanine) was the same as the number of pyrimidine bases (cytosine and thiamine). This was an important fact Watson and Crick had to incorporate into their model of the structure of DNA.

Charles, Jacques Alexandre César (1746–1823) French physicist and physical chemist who, with Gay-Lussac, established a law of the changes in gas volume caused by temperature changes at constant pressure. This is commonly known as Charles' law of pressures.

Jacques Alexandre César Charles

Chevreul, Michel Eugène (1786–1889) French organic chemist who identified the fatty acids oleic acid, butyric acid, capric acid, and caproic acid and a mixture of stearic and palmitic acids called margaric acid. He found that fats consisted of a glycerol "backbone" to which three fatty acids are attached. He discovered hematoxylin, which became an important stain for tissue microscopy, and investigated how a color image could be formed from large numbers of small spots each of a single color – what we now call "pixels."

Chittenden, Russell Henry (1856–1943) American physiologist who discovered the glucose polymer glycogen in muscle and who determined the daily protein requirements of a human being, proving that the then estimate of 118 grams was an overestimate and that 50 grams a day was adequate to maintain health. Chittenden helped to establish biochemistry as a discipline in its own right.

Clausius, Rudolf Julius Emanuel (1822–88) German theoretical physicist who greatly advanced the ideas of Sadi Carnot and Joule, thereby largely establishing thermodynamics. He cleared up previous difficulties by pointing out that heat

Rudolf Julius Emanuel Clausius

cannot pass spontaneously from a cold to a hot body, and furthered the understanding of the kinetic theory of gases. He also promoted the concept of entropy.

Cleve, Per Teodor (1840–1905) Swedish chemist who worked on the rare earths and decided that the "element" didymium, discovered by someone else, was in fact two elements, neodymium and praseodymium. He also discovered holmium and thulium. Ironically, holmium also turned out to be two elements and, in 1886, de Boisbaudran found it was mixed with another new element, dysprosium.

Cohen, Seymour Stanley (b. 1917) American biochemist who in 1946 began to investigate viral infection of cells by tagging viral nucleic acid with radioactive phosphorus. His results strongly suggested that DNA was central to genetics.

Cohen, Stanley H. (b. 1922) American biochemist who worked on DNA-cutting enzymes. He helped to isolate nerve growth factor and went on to isolate epidermal growth factor and to show how this substance interacted with cells to produce a range of effects. Cohen shared the 1986 Nobel Prize in physiology or medicine with Rita Levi-Montalcini.

Corey, Elias James (b. 1928) American chemist who was one of the pioneers in the use of computers to assist in the analysis of methods of synthesis of organic molecules. This has become an indispensable technique and it won Corey the 1990 Nobel Prize in chemistry.

Sir John Cornforth

Cornforth, Sir John (b. 1917) Australian-born British chemist who assisted in the synthesis of penicillin and studied the biosynthesis of cholesterol and various steroids. His most important work, however, was in the detailed elucidation of the mode of action of enzymes, in particular the interaction between an enzyme and its substrate. For this work, he shared the 1975 Nobel Prize in chemistry.

Courtois, Bernard (1777–1838) French chemist who, working with Gouton de Morveau, isolated morphine from opium and later accidentally discovered the element iodine.

Cram, Donald J. (b. 1919) American chemist who in 1972 described the synthesis of left- and right-hand (chiral) structural forms of the molecules of certain cyclic polyethers (crown ethers) and achieved a method of the separation of left- and right-handed forms (enantiomers) so as to produce enantiometrically pure samples. This is important because enantiomers may have different biological properties. For this work, he shared the 1987 Nobel Prize in chemistry.

Crawford, Adair (1748–95) Irish physician and chemist who suggested that animal heat is distributed throughout the body by the arterial blood.

Crick, Francis Harry Compton (b. 1916) English molecular biologist who with Watson in 1953 built a molecular model of the complex genetic material deoxyribonucleic acid (DNA). Crick was the principal solver of the riddle of the genetic code, showing that different triplets of bases defined different amino acids in the protein sequence. With Watson and Wilkins, he was awarded the Nobel Prize in physiology or medicine in 1962. Later, Crick turned to neurophysiology and studied the functioning of the brain.

Cronstedt, Axel F. (1722–65) Swedish mineralogist who in 1751 was the first to isolate nickel. He then demonstrated its magnetic properties. Cronstedt wrote an influential work, *Essay towards a System of Mineralogy* (1758), in which he suggested that minerals should be classified using their chemical composition.

Crutzen, Paul (b. 1933) Dutch chemist who, working with Rowland and Molina, alerted the world to the danger of damage being caused to the ozone layer of the atmosphere, about 9–30 miles up, from artificial nitrogen oxides and chlorofluorocarbon (CFC) gases. For this work, the three men were awarded the 1995 Nobel Prize in chemistry.

Curie, Marie (née Sklodowska) (1867–1934) Polish-born French physicist and wife of Pierre Curie, with whom she worked on magnetism and radioactivity, a term she invented in 1898. Her work on radioactivity earned her the Nobel Prize in physics in 1903. She isolated polonium and, in 1910, pure radium. For this work, she was awarded the Nobel Prize in chemistry in 1911. She died from leukemia, a martyr to long exposure to radioactivity.

Marie Curie

Pierre Curie

Curie, Pierre (1859–1906) French physicist, husband of Marie Curie who, in 1880, discovered piezoelectricity, the property of certain crystals to deform slightly in an electric field and to produce such a field if deformed. Without piezoelectricity, personal computers, ultrasound scanners and cheap phonograph pickups would have been impossible. He also showed that ferromagnetic materials lose their magnetism at certain temperatures (the Curie point). Curie worked with his wife on radioactivity and showed that emitted particles were either electrically negative (beta particles), positive (alpha particles), or neutral (gamma rays). Pierre shared the 1903 Nobel Prize in physics with his wife.

Curl, Robert F. (b. 1933) American chemist who, working with Kroto and Smalley, earned the 1996 Nobel Prize in chemistry for the synthesis of an entirely new molecule consisting only of carbon atoms, which was named a buckminsterfullerene (also known as the buckyball molecule) because of the similarity of its structure to the geodesic domes of the architect Buckminster Fuller. The buckyball is a spherical structure of bonded carbon atoms and has many valuable properties, including acting as an efficient lubricant.

Sir Henry Hallett Dale

Daguerre, Louis Jacques Mande (1787–1851) French inventor who discovered that silver iodide, freshly prepared in the dark, was sensitive to light and that an image projected onto a surface covered with this compound could be developed chemically and then fixed to form a permanent photograph. This was the daguerreotype process, the first successful and commercially viable photographic method.

Dale, Sir Henry Hallett (1875–1968) British pharmacologist who isolated acetylcholine from the ergot fungus. In 1921, after a hint from Dale, the German pharmacologist, Loewi, proved that acetylcholine was the neurotransmitter released at nerve endings in the autonomic nervous system. Dale and Loewi shared the 1936 Nobel Prize in physiology or medicine.

John Dalton

Dalton, John (1766–1844) English chemist and teacher whose atomic theory has become the foundation of modern chemistry. His physical research was chiefly on mixed gases; the law of partial pressures is also known as Dalton's law. In 1794, he first described color blindness, which, for a time was known as "Daltonism."

Dam, Carl Peter Henrik (1895–1976) Danish biochemist who discovered vitamin K by showing that a diet deficient in fatty content led to blood-clotting defects in chicks. He and E. A. Doisy were awarded the 1943 Nobel Prize in physiology or medicine for the discovery of the vitamin. Countless babies have been saved from dangerous bleeding by routine administration of vitamin K.

Dana, James Dwight (1813–95) American mineralogist and geologist who classified minerals, coined the term *geosyncline*, studied coral-rock formation, and theorized about the evolution of the Earth's crust. He wrote the first standard reference books in geology and mineralogy.

Daniell, John Frederic (1790–1845) English chemist and meteorologist who invented a hygrometer (1820), a pyrometer (1830), and the Daniell electric cell, or zinc-copper battery (1836). His *Introduction to Chemical Philosophy* was published in 1839. Daniell also used his hygrometer to investigate atmospheric humidity.

Davy, Sir Humphry (1778–1829) English chemist and science propagandist who, through his experiments, discovered the new elements potassium, sodium, barium, strontium, calcium, and magnesium. In 1815, he invented a safety lamp for use in gas-filled coal mines. Faraday worked under Davy and incurred his jealousy and the contempt of Lady Davy.

Sir Humphry Davy

Debye, Peter Joseph William (1884–1966) Dutch-born American physicist and physical chemist who studied molecular structures, the distribution of electric charges in molecules, and the distances between atoms (the turning effect of a force). He studied dielectric constants and developed the theory of dipole moments, showing how these could be applied to understanding the three-dimensional shape of simple molecules, such as those of water. He showed that water molecules were bent and that benzene rings are flat. He received the Nobel Prize in chemistry in 1936 and is remembered in the unit of the dipole moment, the debye.

Peter Joseph William Debye

Deisenhofer, Johann (b. 1943) German molecular biologist who studied the Y-shaped antibody (immunoglobulin) molecule to discover which sites on the molecule served which particular function. With Huber and Michel, he also researched the structure in the purple bacterium *Rhodopseudomonal viridans* in which photosynthesis occurs. This work earned them the 1988 Nobel Prize in chemistry.

D'Elhuyar, Don Fausto (1755–1833) Spanish mineralogist who discovered the element tungsten.

Democritus (c.460–c.370 BCE) Greek philosopher who wrote widely on physics, mathematics, and cosmology. He proposed that all matter consisted of a vast number of tiny particles having a number of basic characteristics, the combinations of which accounted for the variety of objects. This was not a new idea.

Diels, Otto Paul Hermann (1876–1954) German chemist who, working with Alder, discovered a method of synthesizing new cyclic, or ring, organic compounds by heating sterols with selenium to produce steroids. Diels and Alder were awarded the Nobel Prize in chemistry in 1950.

Gerhard Domagk

Domagk, Gerhard (1895–1964) German physician who showed that the red dye prontosil could kill certain bacteria in animals. During his research, his own daughter became gravely ill with a streptococcal infection. In desperation, Domagk injected her with the dye and she made a full recovery. Domagk's work led to the development of the sulfa drugs. In 1947, he was awarded the Nobel Prize in physiology or medicine.

Dorn, Friedrich Ernst (1848–1916) German chemist who discovered the chemically almost inert, but medically dangerous, radioactive gaseous element radon, a noble gas, and showed that it arose as a decay product of radium.

DuFay, Charles (1698–1739) French chemist whose main contribution to science was in physics. He studied and described the properties of magnetism, showed how magnetic field strength varied with distance, and described natural magnetism. He was also interested in static electricity and was the first to show that an electric charge could be positive or negative, that like charges repelled each other, and that unlike charges were mutually attractive.

de Duve, Christian René (b. 1917) English-born Belgian biochemist who used differential centrifugation to separate biochemical tissue fragments into layers and discovered the cell organelles (little organs), the lysosomes, and the peroxisomes. For this work, he shared the 1974 Nobel Prize in physiology or medicine with Albert Claude and George E. Palade.

Christian René de Duve

Edison, Thomas Alva (1847–1939) American inventor and holder of hundreds of patents who invented the phonograph, the tape ticker for notifying stock exchange prices, the carbon granule microphone, the incandescent electric lamp, and the thermionic diode. His example, and that of the economic benefits of technological advance, led to the creation of the modern research laboratory.

Thomas Alva Edison

Ehrlich, Paul (1854–1915) German medical researcher who used aniline dyes for the selective staining of disease organisms and realized that these substances might kill disease germs without killing the patient. He was right, and became the father of chemotherapy. For his studies on immunity he shared the Nobel Prize in medicine or physiology with Elie Metchnikoff in 1908.

Paul Ehrlich

Eigen, Manfred (b. 1927) German physical chemist who carried out research on chemical reactions that occur so quickly that investigation of them is very difficult. Such reactions were previously thought to be instantaneous. He succeeded in developing techniques that allowed the analysis of these reactions and shared with Norrish and George Porter the 1967 Nobel Prize in chemistry.

Ekeberg, Anders G. (1767–1813) Swedish chemist who discovered the metallic element tantalum, so called by him because of the tantalizing difficulty he experienced in persuading its oxide to react with an acid. This makes tantalum useful as a metal.

Elvehjem, Conrad Arnold (1901–62) American biochemist who discovered the cure for the vitamin deficiency disease pellagra. This was the B vitamin nicotinic acid (niacin).

Empedocles (fl. 450 BCE) Greek philosopher who wrote a poem *On Nature* in which he claimed that everything was made of the four elements, earth, air, fire, and water, which either combined or repelled each other. This idea was to hold up the advancement of chemistry for 2000 years. Empedocles is said to have jumped into the crater of the volcano Mount Etna to prove that he was immortal.

Epicurus

Epicurus (341–270 BCE) Greek philosopher who proposed that everything was made from atoms – particles so small that they cannot be subdivided further. The Roman poet Lucretius accepted and described the atomic theory of Epicurus. The Greek word *atom* means "unable to be cut."

Erlenmeyer, Richard August Carl Emil (1825–1909) German chemist who, for his synthetic work, designed the conical Erlenmeyer flask known to all college chemistry students. He synthesized a number of important organic compounds, including guanidine, tyrosine, and isobutyric acid.

Ernst, Richard Robert (b. 1933) Swiss physical chemist whose work on the development and improvement of nuclear magnetic resonance spectroscopy, a powerful technique for determining the molecular structure of organic compounds, won him the Nobel Prize in chemistry in 1991.

Hans von Euler-Chelpin

Eskola, Pentti Elias (1883–1964) Finnish geologist who specialized in metamorphic formations – preexisting rock that has been modified by heat, pressure, or chemical action. In 1915, he asserted that in such rock that has reached chemical equilibrium, the mineral composition is controlled only by the chemical composition. Eskola recognized eight types of metamorphic formation.

von Euler-Chelpin, Hans (1873–1964) Swedish chemist who carried out a great deal of the earliest work on enzymes. He showed their optimum conditions for function, interaction with vitamins, inhibition by metallic ions and other substances, and the distinction between yeast saccharases and those occurring in the intestine. For his work on enzymes he shared the 1929 Nobel Prize in chemistry with Harden.

Michael Faraday

Faraday, Michael (1791–1867) English chemist and physicist, creator of the classical electromagnetic field theory and one of the greatest experimental physicists. He discovered electromagnetic induction (1831) that led to generators, transformers, and electromagnets; proposed the laws of electrolysis (1833); showed the rotation of polarized light by magnetism (1845); and made other fundamental advances.

Feigenbaum, Edward Albert (b. 1936) American computer scientist who has worked on problems of artificial intelligence and expert systems and who has evolved a program that uses mass spectrometer data to identify organic compounds. The program, known as DENDRAL, comes close to achieving a performance as good as that of an expert and knowledgeable chemist.

Fischer, Edmond H. (b. 1920) American biochemist who, working with Edwin Krebs, showed how glucose molecules – the body's main fuel – are released from the storage form, the glucose polymer glycogen. They showed that the enzyme glycogen phosphorylase, which catalyzes the release, is made operative by receiving a phosphate group from ATP and then made nonoperative by losing this group. Fischer and Krebs shared the 1992 Nobel Prize in physiology or medicine.

Fischer, Emil (1852–1919) German organic chemist who discovered the molecular structures of sugars, including glucose, found the structure of purines, isolated and identified amino acids, and worked on the structure of proteins. He received the Nobel Prize in chemistry in 1902 for his work on sugars and purines.

Fischer, Ernst Otto (1918–94) German chemist noted for his elucidation of the structure of the unusual synthetic compound ferrocene, which is a kind of sandwich with carbon rings as the bread and an iron atom as the filling. Thousands of such compounds are now known. For this work, Fischer shared the 1973 Nobel Prize in chemistry with Wilkinson.

Fischer, Hans (1881–1945) German chemist who researched the molecular structure of chlorophyll, determined the structural formulas for biliverdin and bilirubin, synthesized both these bile products, and worked out and synthesized the structure of hemin. For the latter achievement, he was awarded the Nobel Prize in chemistry in 1930.

Fleming, Sir Alexander (1881–1955) Scottish bacteriologist who in 1928 discovered the first antibiotic substance, penicillin, but without isolating it. He also pioneered the use of salvarsan against syphilis, discovered the antiseptic powers of lysozyme

Sir Alexander Fleming

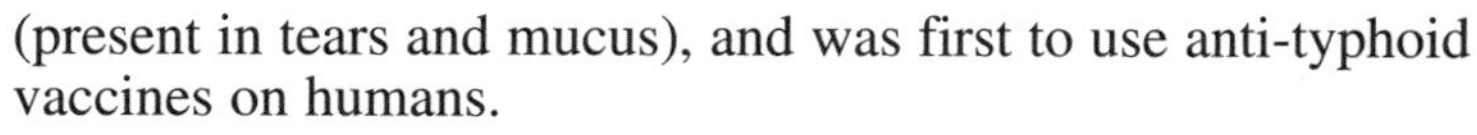

(present in tears and mucus), and was first to use anti-typhoid vaccines on humans.

Howard Florey

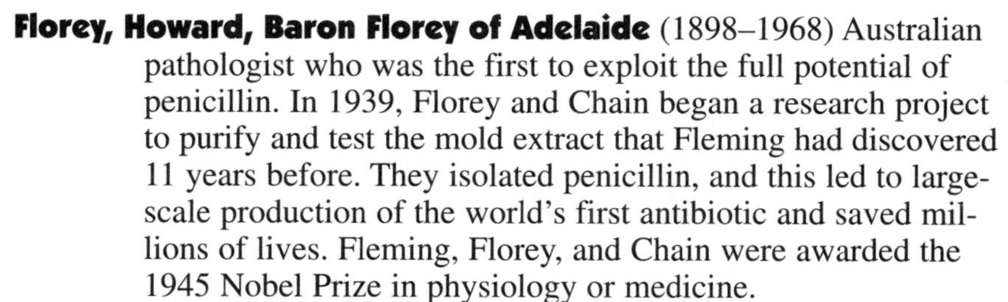

Florey, Howard, Baron Florey of Adelaide (1898–1968) Australian pathologist who was the first to exploit the full potential of penicillin. In 1939, Florey and Chain began a research project to purify and test the mold extract that Fleming had discovered 11 years before. They isolated penicillin, and this led to large-scale production of the world's first antibiotic and saved millions of lives. Fleming, Florey, and Chain were awarded the 1945 Nobel Prize in physiology or medicine.

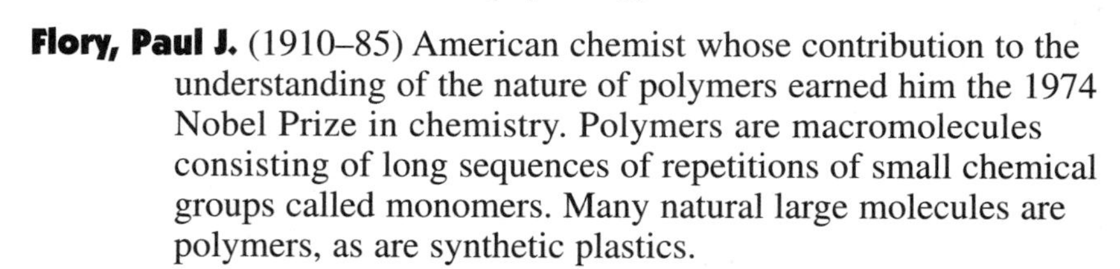

Flory, Paul J. (1910–85) American chemist whose contribution to the understanding of the nature of polymers earned him the 1974 Nobel Prize in chemistry. Polymers are macromolecules consisting of long sequences of repetitions of small chemical groups called monomers. Many natural large molecules are polymers, as are synthetic plastics.

Sir Edward Frankland

Frankland, Sir Edward (1825–99) English organic chemist who became professor at the Royal Institution, London, in 1863. He propounded the theory of valency (1852–60) and, with Joseph Lockyer, discovered helium in the Sun's atmosphere in 1868.

Benjamin Franklin

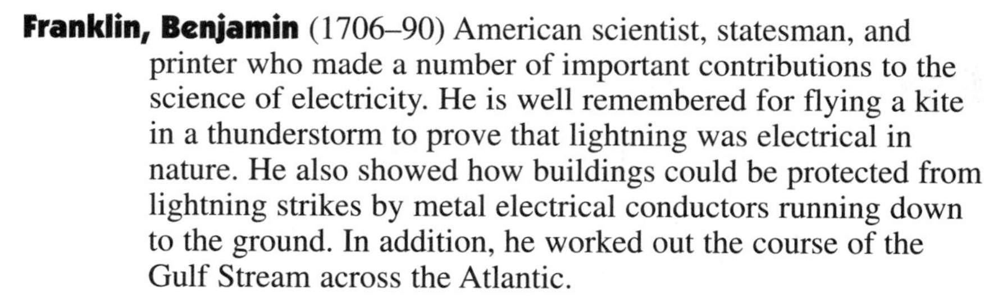

Franklin, Benjamin (1706–90) American scientist, statesman, and printer who made a number of important contributions to the science of electricity. He is well remembered for flying a kite in a thunderstorm to prove that lightning was electrical in nature. He also showed how buildings could be protected from lightning strikes by metal electrical conductors running down to the ground. In addition, he worked out the course of the Gulf Stream across the Atlantic.

Franklin, Rosalind Elsie (1920–58) English X-ray crystallographer whose work, with Wilkins, provided Watson and Crick with the data on which they were able to construct the model of the DNA molecule and achieve scientific immortality. She developed cancer and died in 1958, four years before she could have shared the Nobel Prize with Wilkins, Watson, and Crick.

Fukui, Kenichi (1918–98) Japanese chemist who demonstrated how chemical reactions were essentially a question of the interaction of only two of the orbital electrons of the interacting atoms. Fukui called these "frontier orbitals" and showed how chemical reactions were partly determined by the symmetry of frontier orbitals. He shared the 1981 Nobel Prize in chemistry with Hoffman.

Funk, Casimir (1884–1967) Polish-born U.S. biochemist who isolated the first vitamin and suggested, correctly, that others existed. He believed, wrongly, that they all contained an amine (-NH2) group, and suggested they be called "vital amines" or "vitamines." This was later amended to *vitamins*.

Gassendi, Pierre (1592–1655) French philosopher and astronomer who studied atomism, acoustics, heat, and thermodynamics, and, in his book on the theory of atoms, *Syntagma philosophicum* (1660), introduced the term *molecule* to indicate the smallest unit of a substance capable of an independent existence.

Gay-Lussac, Joseph Louis (1778–1850) French chemist and physicist who investigated the expansion of gases with rising temperature and independently formulated the law known as Charles' law of pressures. He investigated the laws of the combination of gases and the properties of many chemical compounds and elements, especially the halogens. He also compiled charts of the solubility of many compounds.

Giauque, William F. (1895–1982) Canadian-born American chemist who researched the properties of matter at very low temperatures and, by a magnetic process, achieved temperatures within one degree of absolute zero. He also developed a thermometer for measuring very low temperatures. He was awarded the 1949 Nobel Prize in chemistry.

William F. Giauque

Gibbs, Josiah Willard (1839–1903) American physical chemist whose theory of chemical thermodynamics became the foundation of physical chemistry. He also did pioneering work in statistical mechanics, the reports of which were submerged in difficult papers in obscure journals until after they had been independently repeated by scientists of the caliber of Planck and Einstein.

Gilman, Alfred G. (b. 1941) American biochemist who, working with Rodbell, discovered the G protein, a class of chemical messengers that transfer incoming information from receptors in cell membranes to the producers of the second messenger – the hormone that then moves to the effector sites within the cell. G proteins remain inactive until a signal reaches the cell. They then activate. Disease processes can interfere with the G proteins. For this work, Gilman and Rodbell were awarded the 1994 Nobel Prize in physiology or medicine.

Johann Rudolf Glauber

Glauber, Johann Rudolf (1604–68) German chemist who discovered hydrated sodium sulfate, which was sold as a laxative under the trade name of Glauber's salt. Glauber also designed improved laboratory equipment that contributed to the production of industrial, agricultural, and medical products.

Thomas Graham

Graham, Thomas (1805–69) Scottish chemist and physicist. He was one of the founders of physical chemistry, and his research on molecular diffusion of gases led him to formulate the law "that the diffusion rate of gases is inversely as the square root of their density." This is known as Graham's law of diffusion.

Gregor, William (1761–1817) English chemist, whose interest in analyzing local soils led him to the discovery of the element titanium, which has since become an important metal for its light weight and resistance to corrosion.

Grignard, Victor (1871–1935) French organic chemist who showed how to combine magnesium with reactive organic halogen compounds to make organomagnesium compounds that could be used to produce alcohols. He also synthesized a further and valuable range of organometallic compounds. These compounds, called Grignard reagents, are very useful in organic synthesis and earned him the 1912 Nobel Prize in chemistry.

Haber, Fritz (1868–1934) German chemist who, with brother-in-law Carl Bosch, invented a process for the synthesis of ammonia from hydrogen and atmospheric nitrogen, thus overcoming the shortage of natural nitrate deposits accessible to the German explosives industry in World War I. Haber's development of the process on an industrial scale provided copious quantities of fertilizers and also prompted the development of the chemical industry and chemical engineering. For his work, Haber received the Nobel Prize in chemistry in 1918.

Otto Hahn

Hahn, Otto (1879–1968) German radiochemist who showed that the radioactive breakdown of certain elements could provide a way of dating some mineral deposits. He is best known, however, for work with slow neutrons, which indicated that it would be possible to initiate and control nuclear fission, nature's most powerful energy source. This led to attempts to harness fission for industrial and military use. He was awarded the Nobel Prize in chemistry in 1944, and it was proposed that element number 108 should be named hahnium in his honor.

Stephen Hales

Hales, Stephen (1677–1761) English botanist and chemist, founder of plant physiology, whose book *Vegetable Staticks* (1727) was the start of our understanding of vegetable physiology. He was one of the first to use instruments to measure the nutrition and movement of liquids within plants. He also invented machines for ventilating, distilling seawater, and preserving meat.

Hall, Charles Martin (1863–1914) American chemist who in 1886 discovered, independently of Paul Héroult, the first economic method of obtaining aluminum from bauxite (electrolytically). He helped to found the Aluminum Company of America and was its vice president from 1890.

Sir Arthur Harden

Hall, Sir James (1761–1832) Scottish geologist who proved the igneous origin of basalt and dolerite rocks by laboratory tests in which he melted and recrystallized minerals. He also showed that molten magma could cause changes in limestone, producing metamorphic rock.

Harden, Sir Arthur (1865–1940) British chemist who shared the 1929 Nobel Prize in chemistry with Euler-Chelpin for their classic work on fermentation enzymes. Harden proved that living organisms were not necessary for fermentation and that the process could be inhibited if a factor was removed by dialysis. This was the enzyme. He also showed that fermentation of sugars started by phosphorylation to form an ester.

Odd Hassel

Hassel, Odd (1897–1981) Norwegian chemist who used X-ray and electron diffraction methods to determine the molecular structure of the petroleum-derived solvent cyclohexane and related compounds. Cyclohexane is not a flat molecule and can adopt a boat-shaped or a chain conformation. Hassel's work brought to the fore the importance of conformational analysis. He was awarded the 1969 Nobel Prize in chemistry.

Hatchett, Charles (c.1765–1847) English chemist who discovered the metallic element columbium (now niobium), and for whom hatchettine, or hatchettite, a yellowish white semitransparent fossil resin or waxlike hydrocarbon found in South Wales coal, was named.

Hauptman, Herbert A. (b. 1917) American chemist who, working with Jerome Karle in the early 1950s, developed a rapid statistical method of using X-ray crystallography to determine the molecular structure of chemical compounds. Their 1953 paper was largely ignored but the method is now fully established. Hauptman and Karle shared the 1985 Nobel Prize in chemistry.

Haworth, Sir Walter N. (1883–1950) English chemist who was the first to establish the molecular structure of vitamin C and who named it ascorbic acid. Haworth shared the 1937 Nobel Prize in chemistry with Karrer.

Johannes Baptista van Helmont

Helmont, Johannes Baptista van (1579–1644) Flemish physician, physiologist, and chemist who invented the word *gas*, derived from a Greek word for "chaos." He distinguished gases other than air; regarded water as a prime element; believed that digestion was due to "ferments" that converted dead food into living flesh; proposed the medical use of alkalis for excess acidity; and believed in alchemy. His works were published by his son.

William Henry

Henry, William (1774–1836) English chemist who showed that the solubility of a gas in a liquid at a given temperature is proportional to its pressure. This is known as Henry's law. He wrote an influential and often reprinted book called *Elements of Experimental Chemistry* (1801).

Herschbach, Dudley (b. 1932) American physical chemist who shared the 1986 Nobel Prize in chemistry with Yuan Tseh Lee and Polyani for his work on the detailed dynamics of chemical reactions. This research was done by a method not previously used in chemistry, in which low-pressure beams of the reacting molecules were made to intersect while sensitive detectors checked for the products.

Herzberg, Gerhard (1904–99) Canadian chemist who used spectroscopic methods to study the energy levels of hydrogen atoms and molecules and of a range of chemical radicals. He was awarded the 1971 Nobel Prize in chemistry.

Hess, Germain Henri (1802–50) Swiss-born Russian chemist who developed thermochemistry and established the law of constant heat summation (Hess's law). This states that the amount of heat evolved in a chemical change is constant whether the reaction occurs in one stage or several.

de Hevesy, George Charles (1885–1966) Hungarian-born Swedish radiochemist who was the first to suggest the use of radioactive tracers in chemical and biological work. This was to become a technique of great power and value. On Bohr's recommendation, Hevesy also searched for and found a new element, number 72, which he named hafnium. He was awarded the Nobel Prize in chemistry in 1943.

Heyrovsky, Jaroslav (1890–1967) Czech chemist who invented the polarograph. This instrument uses a self-cleaning cathode – a narrow tube through which mercury is slowly passed into the solution – and a large nonpolarizable anode. The electrodes are immersed in a dilute solution of the sample and a variable electrical potential is applied to the cell. As each chemical species is reduced at the cathode, the current rises, and the height of each rise is proportional to the concentration of the component. The technique allows detection of trace amounts of metals and allows investigation of ions interacting with the solvent. Heyrovsky was awarded the 1959 Nobel Prize in chemistry.

Higgins, William (1763–1825) Irish chemist who was the first to propose, although without any experimental data, that substances forming chemical compounds do so according to laws of simple and multiple proportions. The principle was later formulated by Dalton.

Sir Cyril Norman Hinshelwood

Hinshelwood, Sir Cyril Norman (1897–1967) English chemist who, simultaneously with Semenov, investigated chemical reaction kinetics in the interwar years, for which they shared the Nobel Prize in chemistry in 1956. He was a linguist and classical scholar, and was president of the Royal Society.

Hisinger, Wilhelm (1766–1852) Swedish mineralogist who discovered the element cerium and published a geological map of southern and central Sweden.

Hjelm, Peter Jacob (1746–1813) Swedish chemist who in 1782 discovered the element molybdenum in a sample of molybdenite sent to him by Scheele, who suspected that a new element might lie therein.

van't Hoff, Jacobus H. (1852–1911) Dutch chemist who pioneered the measurement and study of the rates and mechanisms of chemical reactions under different conditions of temperature, pressure, and so on. He applied thermodynamics to chemical reactions and is considered to be the founder of physical chemistry. He was the first to be awarded the Nobel Prize in chemistry, in 1901.

Hoffmann, Roald (b. 1937) American chemist who, working with Woodward, worked out the rules for the conservation of orbital symmetry during chemical reactions in which ring compounds are formed from chain structures (cyclization), and bonds break and form simultaneously. With Fukui, he was awarded the 1981 Nobel Prize in chemistry for this work, which has proved to be seminal.

August Wilhelm von Hofmann

von Hofmann, August Wilhelm (1818–92) German chemist who obtained aniline from coal products and discovered many other organic compounds, including formaldehyde (1867). He devoted much labor to the theory of chemical types.

Holmes, Arthur (1890–1965) English geologist and geophysicist who put dates to the geological time scale. He determined the ages of rocks by measuring their radioactive constituents and was an early scientific supporter of Wegener's continental drift theory. He was the first to recognize that the Earth's crust formed (solidified) about 4.56 billion years ago. His book *Principles of Physical Geology* (1944) was highly successful.

Hooke, Robert (1635–1703) English physicist, chemist, and architect. One of the most brilliant scientists of his age and one of the most quarrelsome. He formulated Hooke's law of the extension and compression of elastic bodies and effectively invented the quadrant, the microscope, and the first Gregorian telescope. He was curator of experiments at the Royal Society (1662) and later its secretary.

Hopkins, Sir Frederick Gowland (1861–1947) English biochemist who was the first to make a general scientific study of vitamins and to show their importance in the maintenance of health. In 1929, he shared the Nobel Prize in medicine or physiology with Christiaan Eijkmann.

Huber, Robert (b. 1937) German biochemist who helped to determine antibody structure and binding sites; showed that there was a very small structural difference between the active and the inactive form of the enzyme phosphorylase; and, with Michel and Deisenhofer, worked out the detailed structure of the membrane-bound region in which photosynthesis occurs in the purple bacterium *Rhodopseudomonal viridans*. For this work, he and his colleagues were awarded the 1988 Nobel Prize in chemistry.

Irène Joliot-Curie

Joliot-Curie, Irène (1897–1956) French chemist, daughter of Pierre and Marie Curie, who worked with her husband, Jean-Frédéric Joliot-Curie (1900–58), on making the first of a series of artificially produced radioactive isotopes by bombarding aluminum with alpha particles. She and her husband were awarded the 1935 Nobel Prize in chemistry. Like her mother, she died from leukemia, presumably as a result of long exposure to radioactivity.

James Prescott Joule

Joule, James Prescott (1818–89) English physicist who laid the foundations for the theory of the conservation of energy. He is famous for experiments in heat, which he showed to be a form of energy. He also showed that if a gas expands without performing work, its temperature falls. The Joule, a unit of work or energy, is named after him. With Kelvin, he devised an absolute scale of temperature.

Karrer, Paul (1889–1971) Swiss chemist who studied amino acids, proteins, polysaccharides, and plant pigments. He established the molecular structure of carotene and worked on vitamin A and other vitamins. For this work, he shared the 1937 Nobel Prize in chemistry with Norman Haworth.

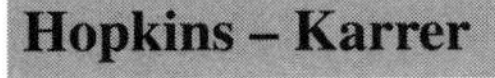

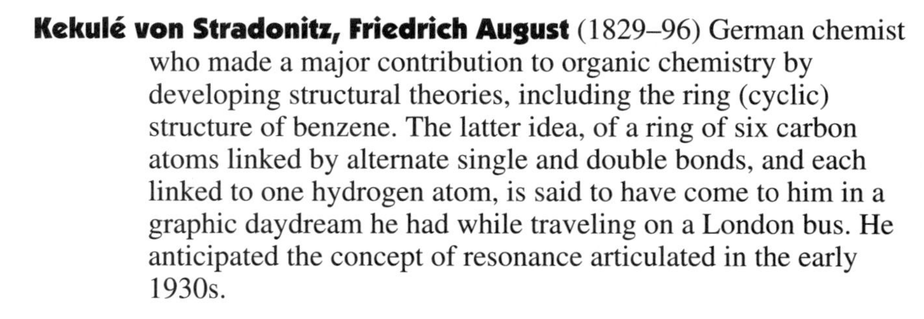
Kekulé von Stradonitz, Friedrich August (1829–96) German chemist who made a major contribution to organic chemistry by developing structural theories, including the ring (cyclic) structure of benzene. The latter idea, of a ring of six carbon atoms linked by alternate single and double bonds, and each linked to one hydrogen atom, is said to have come to him in a graphic daydream he had while traveling on a London bus. He anticipated the concept of resonance articulated in the early 1930s.

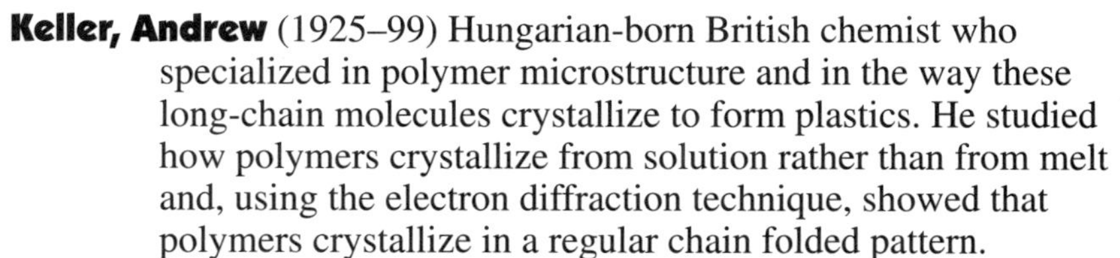
Keller, Andrew (1925–99) Hungarian-born British chemist who specialized in polymer microstructure and in the way these long-chain molecules crystallize to form plastics. He studied how polymers crystallize from solution rather than from melt and, using the electron diffraction technique, showed that polymers crystallize in a regular chain folded pattern.

William Thomson, Lord Kelvin

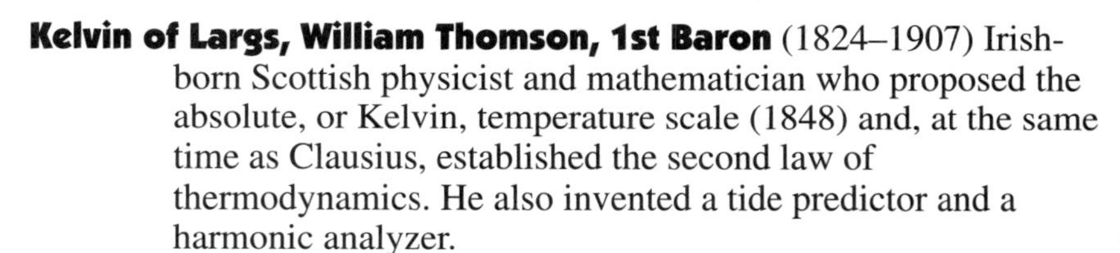
Kelvin of Largs, William Thomson, 1st Baron (1824–1907) Irish-born Scottish physicist and mathematician who proposed the absolute, or Kelvin, temperature scale (1848) and, at the same time as Clausius, established the second law of thermodynamics. He also invented a tide predictor and a harmonic analyzer.

Edward Calvin Kendall

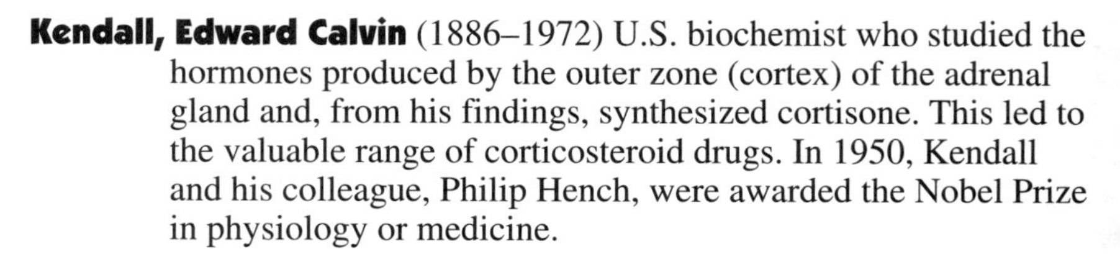
Kendall, Edward Calvin (1886–1972) U.S. biochemist who studied the hormones produced by the outer zone (cortex) of the adrenal gland and, from his findings, synthesized cortisone. This led to the valuable range of corticosteroid drugs. In 1950, Kendall and his colleague, Philip Hench, were awarded the Nobel Prize in physiology or medicine.

Kendrew, Sir John C. (1917–97) English molecular biologist who established the three-dimensional structure of muscle hemoglobin (myoglobin) by X-ray crystallography in 1959. He was awarded the 1962 Nobel Prize in chemistry.

Kirchhoff, Gustav Robert (1824–87) German physicist who, while still a student, derived the laws, now known as Kirchhoff's laws, for determining currents in electrical networks. Working in spectroscopy, he discovered the elements cesium and rubidium, and formulated Kirchhoff's laws of radiation, which stipulate that, for a given wavelength, the ratio of emission to absorption is the same for all bodies at a given temperature.

Klaproth, Martin Heinrich (1743–1817) German chemist who correctly predicted the existence of the elements strontium, titanium, uranium, and zirconium, and confirmed and named tellurium. All these elements were isolated by other chemists. He determined the properties of the minerals yttria and beryllia. Klaproth, who is often considered a founder of analytical chemistry, is noted for his insistence on publishing results that contradicted his own theories.

Klug, Sir Aaron (b. 1926) South African chemist whose work on determining the chemical structure of viruses, including the polio virus, earned him the 1982 Nobel Prize in chemistry. Klug used a variety of techniques, including X-ray diffractions, electron microscopy, and structural modelling.

Kolbe, Adolph Wilhelm Hermann (1818–84) German chemist who was the first to synthesize acetic acid from inorganic material. He discovered the Kolbe reaction in 1859, which allowed large-scale industrial synthesis of salicylic acid. This was important for the production of acetylsalicylic acid (aspirin). He also developed a useful electrolytic process for the synthesis of alkanes.

Adolph Wilhelm Hermann Kolbe

Kosterlitz, Hans Walter (b. 1903) German-born Scottish pharmacologist and physiologist who, working with John Hughes, discovered the natural morphine-like body opiates, the enkephalins (endorphins), and showed that they were blocked by the drug naloxone, which antagonizes morphine.

Krebs, Sir Hans Adolf (1900–81) German-born British biochemist who elucidated the cyclical series of biochemical reactions by means of which food is converted into energy for cell function and for the synthesis of biomolecules. This important process is known as the Krebs cycle and is fundamental to cell physiology. Krebs shared the Nobel Prize in physiology or medicine with Lipmann in 1953.

Sir Hans Adolf Krebs

Kroto, Sir Harold (b. 1939) English chemist who, with Smalley and Curl, succeeded in making an entirely new molecule, consisting only of carbon atoms and named a buckminsterfullerene ("buckyball") because of the similarity of its structure to the geodesic domes of the architect Buckminster Fuller. The buckyball is a spherical structure of bonded carbon atoms and has many valuable properties. The three men were awarded the 1996 Nobel Prize in chemistry.

Kuhn, Richard (1900–67) German chemist who worked on enzymes and the three-dimensional structure of molecules (stereochemistry) and who made important discoveries in the field of vitamins, especially vitamin A. He was awarded the 1938 Nobel Prize in chemistry.

Antoine Laurent Lavoisier

Lavoisier, Antoine Laurent (1743–94) French chemist who has been described as the founder of modern chemistry. He discovered and named oxygen and proved its importance in respiration, combustion, and rusting and as an element that formed many compounds (oxides) with metals. His book *Traité Elémentaire de Chimie*, published in 1789, described his experiments. He was guillotined during the Reign of Terror.

Le Chatelier, Henri Louis (1850–1936) French chemist noted for his rule, known as Le Chatelier's principle, which states that every change in a system in stable chemical equilibrium results in a rearrangement of the system so that the original change is minimized. He devised a railway water brake, an optical pyrometer, and contributed to the field of metallurgy.

Leclanché, Georges (1839–82) French engineer who devised an electric cell using a zinc rod and a porous earthenware pot containing a carbon rod surounded by manganese dioxide and carbon black all, enclosed in a jar filled with a solution of sal ammoniac. This highly successful cell was later converted to a dry form and was used as a source of portable electric energy throughout most of the 20th century.

Yuan Tseh Lee

Lee, Yuan Tseh (b. 1936) American physical chemist who used molecular beam techniques to study the dynamics of chemical reactions. Lee is credited with a major contribution to the success of developing the molecular beam method and shared the 1986 Nobel Prize in chemistry with Polanyi and Herschbach.

Lehn, Jean-Marie (b. 1939) French chemist who demonstrated that sodium and potassium ions can pass across biological membranes in a nonpolar environment by being enclosed within a cavity or channel in a large organic molecule. This discovery opened up a new branch of organic chemistry, called supramolecular chemistry, and it won Lehn a share of the 1987 Nobel Prize in chemistry with Pedersen and Cram.

Leloir, Luis F. (1906–87) Argentinian biochemist who made a number of biochemical advances important to medicine. He discovered the hormone angiotensin, which raises blood pressure; he showed how the energy-storage polysaccharide glycogen is a polymer built up from units of glucose; and he showed how galactose was converted to glucose. For these findings, he was awarded the 1970 Nobel Prize in chemistry.

Leucippus (fl. 500 BCE) Greek philosopher who is said to have originated the atomistic theory that was taken up by Democritus and the poet Lucretius. The term *atom* derives from the Greek *a*, and *tomos*, "a cut," implying an entity so small that it cannot be subdivided further.

Lewis, Gilbert Newton (1875–1946) American physical chemist who developed theories on chemical thermodynamics, atomic structure, and atomic bonding. He pioneered work on the electronic theory of valency, showing the difference between ionic and covalent bonds. He defined an acid as an electron acceptor and a base as an electron donor.

Li, Choh Hao (1913–89) Chinese-born American biochemist who isolated several pituitary hormones and worked out the amino acid sequence of growth hormone and then synthesized it. He also established the sequence in ACTH, the pituitary hormone that prompts the adrenals to produce cortisone.

Libby, Willard Frank (1908–80) American chemist noted for developing the method of radiocarbon dating for determining the age of once living organic material. For this important advance he was awarded the Nobel Prize in chemistry in 1960.

Liebig, Justus, Freiherr von (1803–73) German chemist and one of the most illustrious chemists of his age, equally great in method and in practical applications. He made his name both in organic and animal chemistry, and in the study of alcohols. He was the founder of agricultural chemistry and the discoverer of chloroform.

Lipmann, Fritz Albert (1899–1986) German-born American biochemist who showed how citric acid is formed from

Justus Liebig

Fritz Albert Lipmann

oxaloacetate and acetate and that an unrecognized cofactor, coenzyme A, was required. Lipmann isolated this factor. The formation of citric acid is the first step in the important energy-producing Krebs cycle. Lipmann shared the 1953 Nobel Prize in physiology or medicine with Krebs.

Lipscomb, William N. (b. 1919) American chemist who used novel and ingenious methods to work out the molecular structure of a number of boron compounds using X-ray crystal diffraction methods at low temperatures. His methods became more generally useful and he was awarded the 1976 Nobel Prize in chemistry.

Loewi, Otto (1873–1961) German pharmacologist who proved that the nerve impulse was transferred from nerve to muscle by a chemical mediator. He distinguished acetylcholine from adrenaline for this function, and the former was later identified by Dale. He shared the Nobel Prize in physiology or medicine with Dale in 1936.

Lomonosov, Mikhail (1711–65) Russian scientist who set up the country's first chemical laboratory. He seems to have been well ahead of his time and is said to have proposed the law of conservation of mass, the wave theory of light, and the kinetic theory of heat well before these important principles were understood in the West. He believed in popular education.

Lucretius (fl. c.100 BCE) Roman poet and philosopher who wrote a long poem called *De rerum natura* (*On the Nature of Things*) in which he outlined the atomist theory. He described a large army, seen from a remote cliff top, which appeared to be a solid body but which reason confirmed to be composed of individual particles.

Marcus, Rudolph A. (b. 1923) Canadian-born American chemist whose work on the theory of electron transfer in chemical reactions, such as oxidation and reduction, changed the way scientists looked at these reactions and provided a clearer understanding of a wide range of chemical processes. This important work earned him the 1992 Nobel Prize in chemistry.

Martin, Archer J. P. (b. 1910) English biochemist who developed methods of partition chromatography, using columns of silica gel, for the separation of amino acids from the mixture produced by hydrolysis of proteins. This method greatly facilitated the work of determining the structure of proteins. He was awarded the 1952 Nobel Prize in chemistry.

Meitner, Lise (1878–1968) Austrian nuclear physicist who discovered the radioactive element protoactinium. She made the telling, and correct, suggestion that the presence of radioactive barium in the products of uranium, which had been bombarded with neutrons, was probably due to the fact that some uranium atom nuclei had been split in two. The transuranic element number 108, meitnerium, has been named for her.

Lise Meitner

Mendeleyev, Dmitri Ivanovich (1834–1907) Russian chemist who arranged the known elements into a table of columns by their chemical properties. This was the periodic table, which was of great importance in the development of chemistry and enabled Mendeleyev to predict the existence of several elements that were subsequently discovered. Element 101 is named mendelevium after him. He also worked on the liquifaction of gases, the expansion of gases, and a theory of solutions.

Merrifield, Bruce (b. 1921) American chemist who developed an ingenious and rapid way of synthesizing proteins by lining up the constituent amino acids in the right order on a polystyrene bead, a process that has now been automated. This work earned him the 1984 Nobel Prize in chemistry.

Bruce Merrifield

Meyerhof, Otto Fritz (1884–1951) German-born American biochemist whose work on muscle physiology showed that lactic acid was produced from muscle glycogen during muscle contraction in anaerobic conditions. He also showed that the utilization of glucose as a fuel in living cells involved a cyclic biochemical pathway. For these discoveries, he received the Nobel Prize in physiology or medicine in 1922.

Michel, Hartmut (b. 1948) German biochemist who, working with Huber and Deisenhofer, established the structure of an area of a bacterium in which photosynthesis takes place. He and his colleagues were awarded the 1988 Nobel Prize in chemistry for this work.

Miller, Stanley (b. 1930) American chemist who studied the possible origins of life on Earth by using laboratory equipment to simulate supposed early atmospheric gaseous content and electric sparks to simulate lightning. He succeeded in forming amino acids, the units of proteins. Later work on the enzymatic function of RNA added credibility to Miller's ideas.

Mitchell, Peter (1920–92) English biochemist who revolutionized thought on the process of oxidative phosphorylation in which adenosine triphosphate (ATP) is regenerated from adenosine diphosphate (ADP) and phosphate. Breakdown of ATP to ADP releases large amounts of energy for cell functions from the phosphate bonds. Mitchell proposed that electron transport formed a proton gradient across the mitochondrial membrane that directly brought about the synthesis of ATP from ADP. He was awarded the 1978 Nobel Prize in chemistry.

Mohs, Friedrich (1773–1839) German mineralogist who wrote *The Natural History System of Mineralogy* (1821) and *Treatise on Mineralogy* (1825). He classified minerals on the basis of hardness, and the Mohs' scale of hardness is still in use. This scale, 0–10, is based on the ability of any mineral to scratch another lower down the scale. Talc is 1, diamond is 10.

Henri Moissan

Moissan, Henri (1852–1907) French chemist who isolated the element fluorine in 1886. He invented an electric arc furnace with which he achieved very high temperatures and synthesized rubies. He was the founder of high-temperature chemistry. He was awarded the 1906 Nobel Prize in chemistry for his work on fluorine.

Molina, Mario (b. 1943) American chemist who, working with Rowland and Crutzen, alerted the world to the danger of damage being caused to the ozone layer of the atmosphere, about 9–30 miles up, from artificially produced nitrogen oxides and chlorofluorocarbon (CFC) gases. The ozone layer protects us against dangerous concentrations of ultraviolet frequencies in sunlight. For this work, the three men were awarded the 1995 Nobel Prize in chemistry.

Stanford Moore

Monod, Jacques Lucien (1910–76) French biochemist who worked with François Jacob on messenger RNA. In *Chance and Necessity* (1970) he proposed that humans are the product of chance in the universe.

Moore, Stanford (1913–82) American biochemist who, working with Stein, invented a chromatography process that could separate from a mass of enzyme-digested protein all the constituent amino acids so that they could be identified and quantified. Moore and Stein also invented an automated method of determining the base sequence in a length of RNA. Moore, Stein and Anfinsen were awarded the 1972 Nobel Prize in chemistry.

Morley, Edward William (1838–1923) American chemist and physicist who, with Albert Michelson, developed a sensitive interferometer with which they showed (1887) that the speed of light is constant, whether measured in the direction of the Earth's movement or perpendicular to that direction (Michelson-Morley experiment).

Moseley, Henry Gwyn Jeffreys (1887–1915) British physicist who used X rays' scattering by different elements to show that the resulting wavelengths decreased regularly with increase of atomic weight. He concluded, correctly, that each element had a different number of electrons.

Mulliken, Robert S. (1896–1986) American chemist who, in the 1930s, helped to develop the molecular orbital theory of molecular structure and chemical bonding. He was awarded the 1966 Nobel Prize in chemistry.

Robert S. Mulliken

Mullis, Kary B. (b.1944) American biochemist who, while driving one evening, conceived the idea of the polymerase chain reaction (PCR) that was to become one of the most important advances in genetic research, engineering, and medicine since Crick and Watson. PCR provides millions of copies of any DNA fragment. He was awarded the Nobel Prize in chemistry in 1993.

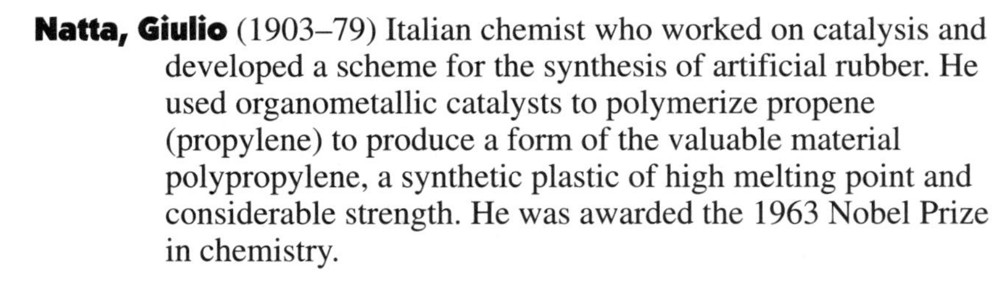

Natta, Giulio (1903–79) Italian chemist who worked on catalysis and developed a scheme for the synthesis of artificial rubber. He used organometallic catalysts to polymerize propene (propylene) to produce a form of the valuable material polypropylene, a synthetic plastic of high melting point and considerable strength. He was awarded the 1963 Nobel Prize in chemistry.

Walther H. Nernst

Nernst, Walther H. (1864–1941) German chemist who discovered the third law of thermodynamics, that entropy (a measure of the unavailability of a system's energy to do work) increases as temperature approaches absolute zero. He worked on electrical insulators, and the specific heat of solids at low temperature, and explained the chain reaction by which chlorine and hydrogen explode on exposure to light. He was awarded the 1920 Nobel Prize in chemistry.

Newlands, John Alexander Reina (1836–98) British chemist who, like Mendeleyev, was one of the first to show that the properties of chemical elements changed in a periodic manner. He arranged the then known 62 elements in order of increasing atomic weight and showed that these could be placed into groups of eight based on similar properties. This was known as the law of octaves.

Nicholson, William (1753–1815) English chemist who showed that water could be broken down into hydrogen and oxygen by inserting two wires into it that were connected to an electric battery. This was the first demonstration of electrolysis. Nicholson, who had had the benefit of publishing his own scientific journal, was able to report some findings with the voltaic pile even before Alessandro Volta.

Alfred Bernhard Nobel

Nobel, Alfred Bernhard (1833–96) Swedish chemist and engineer who discovered the element nobelium and invented a safer explosive by mixing nitroglycerine with the diatomaceous earth kieselguhr. He called this explosive dynamite. It earned him so much money that he was able to leave more than $9 million for the establishment of the Nobel Prize in five fields.

Noddack, Walter (1893–1960) German chemist who, working with his wife, Ida, discovered the elements rhenium and technetium, and did research on the photopigments of the eye.

Norrish, Ronald G. W. (1879–1978) British chemist who was one of the pioneers of photochemistry, and who invented flash photolysis, a technique in which photochemical change caused by a very brief, bright flash of light can immediately be studied by the absorption spectra of the resultant materials. For this advance, he was awarded the 1967 Nobel Prize in chemistry, together with Eigen and George Porter.

Northrop, John H. (1891–1987) American chemist who crystallized the protein-digesting enzyme pepsin and showed it to be a protein. He then worked on other large molecules and on the purification of enzymes. He isolated the first virus infecting a bacterium. He was awarded the 1946 Nobel Prize in chemistry.

Ochoa, Severo (1905–93) Spanish-born U.S. biochemist who became a professor of biochemistry at New York University in 1954. A year later, he showed how cells use an enzyme to join fragments of DNA. This work led to genetic engineering. For this discovery, Ochoa and a fellow worker on DNA, Arthur Kornberg, shared the 1959 Nobel prize in physiology or medicine.

Olah, George A. (b. 1927) American organic chemist who found a way, using powerful acids, of extending the life of fragments of hydrocarbon molecules – compounds that appeared only momentarily as intermediate stages in chemical reactions. This work threw important light on the details of chemical reactions and won Olah the 1994 Nobel Prize in chemistry.

Onsager, Lars (1903–76) Norwegian-born American chemist whose work on the thermodynamics of irreversible processes earned him the 1968 Nobel Prize in chemistry. He also worked on strong electrolytes.

Wilhelm Ostwald

Ostwald, Wilhelm (1853–1932) German chemist who was the pioneer of modern physical chemistry and who showed how catalysts work. He was awarded the 1909 Nobel Prize in chemistry.

Pasteur, Louis (1822–95) French chemist and founder of modern bacteriology who proposed the "germ" theory of disease in the late 1860s. This was, perhaps, the greatest single advance in the history of medicine. He also developed pasteurization: rapid, short-term heating to kill harmful bacteria in wine and milk.

Pauling, Linus Carl (1901–94) American chemist noted for his germinal work *The Nature of the Chemical Bond* (1939), which applied quantum theory. He also made major advances in the understanding of protein structures and was awarded the Nobel Prize in chemistry in 1954 for his contributions to the electrochemical theory of valency. His work also covered inorganic complexes, protein structure, antibodies, DNA structure, and the molecular basis of some genetic diseases. His belief in the medical efficacy of vitamin C was condemned during his lifetime but has now been vindicated with the recent understanding of the biological effect of free radicals and the value of vitamin C as a biological antioxidant.

Pedersen, Charles J. (1904–90) American chemist who produced a crown-shaped cyclic polyether that was given the name "crown ether," and discovered that compounds of this kind would bind sodium and potassium ions strongly, making alkali metal salts that were soluble in organic solvents. This work helped to explain how these metallic ions were transported across biological membranes, a matter of great importance in physiology and pharmacology. For this work, he was awarded a share in the 1987 Nobel Prize in chemistry.

Pelletier, Pierre Joseph (1788–1842) French chemist who named the green leaf pigment chlorophyll, and worked on alkaloids, isolating many, some of which have become important in medicine.

Sir William Henry Perkin

Perkin, Sir William Henry (1838–1907) English chemist who worked as assistant to Hofmann and, in 1856, discovered a brilliant purple dye. Later named mauveine, his invention became immensely popular, earned him a fortune, and led to the foundation of the modern synthetic dye industry. Some of his work on the synthesis of organic compounds, known as the Perkin synthesis, led to the development of the synthetic perfume industry.

Max F. Perutz

Perutz, Max F. (b. 1914) Austrian-born British biochemist who, using X-ray diffraction and other methods, achieved the extraordinarily complex task of determining the molecular three-dimensional structure of hemoglobin. For this work, he shared the 1962 Nobel Prize in chemistry.

Polanyi, John C. (b. 1929) Canadian physical chemist who studied the infra-red light emitted during chemical reactions. This provided information about the distribution of energy in molecules and won him the 1986 Nobel Prize in chemistry.

Porter, George (Baron Porter of Luddenham) (b. 1920) British physical chemist who researched photochemistry and, with Norrish, developed the technique of flash photolysis. He, Norrish and Eigen shared the 1967 Nobel Prize in chemistry.

Porter, Rodney Robert (1917–85) British biochemist who first suggested that antibodies were Y-shaped. In 1962, Porter showed that the gamma globulin antibody could be split by an enzyme into three large fragments. Two of these could bind antigens and were known as "Fab" (fragment antigen binding); the third, a crystalline fragment, could not. Porter showed that this third fragment was common to all antibodies, and that it was the Fab fragments that existed in thousands of different forms that give antibodies their specificity. This important discovery led to his being awarded the 1972 Nobel Prize in physiology or medicine.

Pregl, Fritz (1869–1930) Austrian chemist who developed new techniques of microanalysis to study the tiny quantities of biochemical substances available to him in his work. He developed a weighing balance of unprecedented accuracy and sensitivity and was awarded the 1923 Nobel Prize in chemistry.

Prelog, Vladimir (1906–98) Yugoslavian-born Swiss chemist who synthesized adamantine, a molecule related to diamond. His main work was in the study of molecular shape and, in particular, those molecules that could exist in two shapes, one being the mirror image of the other (chirality). He went on to study the three-dimensional structure of molecules (stereochemistry) in general and shared the 1975 Nobel Prize in chemistry for his work on the stereochemistry of enzymes.

Joseph Priestley

Priestley, Joseph (1733–1804) English chemist and Presbyterian minister who pioneered the study of the chemistry of gases, and, in 1774, was one of the discoverers of oxygen. He was not the first to identify oxygen, as is often stated, but he earned recognition through publication. The Swedish apothecary Scheele isolated oxygen in 1772.

Prigogine, Ilya (b. 1917) Russian-born Belgian chemist who researched the thermodynamics of irreversible chemical processes and learned how to handle processes far from equilibrium. For this work, he was awarded the 1977 Nobel Prize in chemistry.

Joseph Louis Proust

Proust, Joseph Louis (1754–1826) French chemist who studied how elements combined to form molecules and formulated the law of definite proportions, which states that regardless of the way a compound is prepared, it always contains the same elements in the same proportions.

Prusiner, Stanley B. (b. 1942) American neurologist and biochemist who was the first scientist since the discovery of viruses to detect an entirely new infective agent. Prusiner, as a young neurology resident, was in charge of a patient who died of Creutzfeldt-Jakob disease (CJD). He decided to research the cause. Ten years later, he isolated small protein bodies, which he called prions, and showed that these were the cause of CJD and of the similar bovine spongiform encephalopathy. He was awarded the Nobel Prize in physiology or medicine in 1997.

Sir William Ramsay

Ramsay, Sir William (1852–1916) Scottish physical chemist who isolated the five elements argon, neon, krypton, xenon, and radon, constituting the whole class of the noble, or inert, gases in the periodic table. He was also the first to isolate helium, previously believed to exist only in the Sun. His writings include *The Gases of the Atmosphere* and *Elements and Electrons*. He was awarded the Nobel Prize in chemistry in 1904. Radon's atomic weight was determined by Ramsay.

Rhazes

Reichstein, Tadeus (1897–1976) Polish-born Swiss biochemist whose work led to the synthesis of vitamin C and an understanding of the chemistry of the natural corticosteroid hormones of the adrenal gland. He was able to isolate 29 natural steroids. This work led to the production of a range of steroid drugs of great medical value that have saved many lives. In 1950, he shared the Nobel Prize in physiology or medicine.

Rhazes (Al-Razi) (854–925) Persian physician and alchemist who based his practice on rational grounds, observation, and experience; taught high ethical standards in medical care; and treated poor patients without fees. He recorded all the medical knowledge of his time and wrote 10 medical treatises himself.

Richards, Theodore William (1868–1928) American chemist who determined with great accuracy the atomic weights of 25 elements. This led to the discovery of natural isotopes of elements, each of which has a slightly different weight because of the different number of neutrons. In 1905, he introduced the adiabatic calorimeter, an instrument that measured the heat rise from combustion without error from loss or gain of heat. He received the Nobel Prize in chemistry in 1914.

Richter, Hieronymous Theodor (1824–98) German chemist who worked with Ferdinand Reich and discovered the metallic element indium.

Robinson, Sir Robert (1886–1975) British chemist who helped to develop penicillin production; conducted research into alkaloids; studied dyes; produced a theory of cyclic benzene-like compounds (aromaticity); and worked on the chemistry of natural products, for which he was awarded the 1947 Nobel Prize in chemistry.

Rodbell, Martin (b. 1925) American biochemist who, working with Gilman, discovered the G protein, a previously unknown class of chemical messengers that, activated by an external hormone (the "first messenger") binding to a cell membrane receptor, effectively turn on the producers of the "second messenger" – the hormone that then moves to the effector sites within the cell and initiates the effect of the external hormone. Disease processes can interfere with the G proteins. For this work, Rodbell and Gilman were awarded the 1994 Nobel Prize in physiology or medicine.

Rose, William Cumming (1887–1984) American biochemist who investigated the individual role of the 20 amino acids in dietary protein and discovered that 10 of them were indispensable to rats but only 8 were indispensable to humans. These are known as the "essential" amino acids. The others can be synthesized in the body.

Rowland, F. Sherwood (b. 1927) American chemist who, working with Crutzen and Molina, alerted the world to the danger of damage being caused to the ozone layer of the atmosphere, about 9–30 miles up, from artificially produced nitrogen oxides and chlorofluorocarbon (CFC) gases. For this work, they were awarded the 1995 Nobel Prize in chemistry.

Rutherford, Daniel (1749–1819) Scottish chemist who was one of the discoverers of nitrogen gas.

Ruzicka, Leopold (1887–1976) Swiss chemist whose study of perfumes led to a detailed investigation of the unsaturated hydrocarbon essential oils, multi-membered ring structures known as the terpenes. He then discovered their structural relation to the steroids. With Butenandt, he was awarded the 1939 Nobel Prize in chemistry. His wealth from his discoveries enabled him to set up an art gallery of Dutch and Flemish masters.

Paul Sabatier

Sabatier, Paul (1854–1941) French chemist whose most important work was in the catalyzed hydrogenation of unsaturated organic compounds. This found wide industrial applications, the best known of which is the hydrogenation of vegetable oils to produce margarine and other butter substitutes. He was awarded the 1912 Nobel Prize in chemistry.

Julius von Sachs

von Sachs, Julius (1832–97) German botanist who proved that chlorophyll is critical in the natural synthesis of sugars from carbon dioxide and water and that oxygen was released. He was also the first to find chlorophyll in plant chloroplasts.

Sanger, Frederick (b. 1918) English biochemist who after 12 years of work was able to establish the molecular structure of the protein insulin with its 51 amino acids. He was also able to show the small differences between the insulins of different mammals. Sanger later turned to DNA sequencing and, by laborious methods, was able to determine the base sequence of mitochondrial DNA and of the whole genome of a virus. For his work on insulin, he was awarded the Nobel Prize in chemistry in 1958, and for his achievements in DNA sequencing, he was awarded the 1980 Nobel Prize in chemistry.

Scheele, Carl Wilhelm (1742–86) Seriously undervalued Swedish apothecary and chemist who was the actual discoverer of oxygen and nitrogen as well as of the elements arsenic, barium, chlorine, manganese, and molybdenum. He was the victim of scientific neglect by his contemporaries and, for a long time, by scientific historians, but is now being recognized as the chemistry genius that he was. Scheele's life may have been shortened by his habit of tasting every new substance he discovered.

Glenn T. Seaborg

Seaborg, Glenn T. (1912–99) American chemist and atomic scientist who discovered many previously unknown isotopes of common elements. He assisted in the production of a number of non-natural, above-uranium (transuranic) elements, including neptunium (93), plutonium (94), americium (95), berkelium (97), einstinium (99), fermium (100), and nobelium (102). During his lifetime, he was honored by having element 106 named seaborgium. He was also involved in the production of the fissionable isotope plutonium-239, which has formed the basis of atomic weapons ever since. He was awarded the 1951 Nobel Prize in chemistry.

Sefström, Nils G. (1765–1829) Swedish chemist who in 1880 discovered vanadium, the metallic element later alloyed with steel to produce very high-strength, low-corrosion metal for tools and other purposes. Vanadium was actually discovered by Andres del Rio in 1801, but he let himself be persuaded that the substance he had found was an impure form of chromium.

Semenov, Nikolai Nikolaevich (1896–1986) Russian physical chemist who made notable contributions to chemical kinetics, especially those of chemical chain reactions. He also studied the features of combustion and explosions. For this work, he was awarded the Nobel Prize in physics in 1956, the first Soviet citizen to achieve this distinction.

Nikolai Nikolaevich Semenov

Skou, Jens C. (b. 1918) Danish chemist who established that the enzyme sodium, potassium-ATPase was the first enzyme known to promote the transport of ions across a cell membrane. Ionic transport across membranes is fundamental to the transmission of nerve impulses. For this work, Skou was awarded the Nobel Prize in chemistry in 1997.

Smalley, Richard E. (b. 1943) American chemist who shared the Nobel Prize in chemistry in 1996 with Kroto and Curl for the synthesis of an entirely new molecule, consisting only of carbon atoms, named a buckminsterfullerene ("buckyball") because of the similarity of its structure to the geodesic domes of the architect Buckminster Fuller. The buckyball is a spherical structure of bonded carbon atoms and has many valuable properties.

Michael Smith

Smith, Michael (b. 1932) British-born Canadian chemist who discovered how to produce deliberate mutations in DNA at precise locations (site-directed mutagenesis), a technique that enabled him to code for new proteins with new properties. He shared the 1993 Nobel Prize in chemistry with Mullis.

Frederick Soddy

Soddy, Frederick (1877–1956) English radio chemist who worked with the physicist Ernest Rutherford on the development of the general theory of the decay of radioactive elements. They showed that this occurred because of the emission of alpha and beta particles and gamma radiation. In 1903, Soddy, working with Ramsay, discovered that helium was formed during radioactive decay. Alpha particles are helium nuclei and immediately acquire electrons. Soddy also discovered that a radioactive element may have several atomic weights and coined the term *isotope* to indicate that they all occupied the same place in the periodic table. This is because they all have the same number of protons, hence electrons, and hence the same chemical properties. Soddy was awarded the 1921 Nobel Prize in chemistry.

Sørensen, Søren Peter Lauritz (1868–1939) Danish chemist who in 1909, while describing the effect of hydrogen ion concentration on enzyme activity, proposed the use of a negative logarithm of this concentration as a measure of acidity and alkalinity. This became the standard pH scale now in universal use. He also studied amino acids, enzymes, and proteins. He and his wife were the first to crystallize the egg protein albumin.

Hermann Staudinger

Staudinger, Hermann (1881–1965) German chemist who invented an improved way of synthesizing isoprene, the structural unit of natural and synthetic rubber. He insisted, correctly, against opposition, that rubber was a polymer macromolecule and researched biological polymers. His work on polymer chemistry won him the 1953 Nobel Prize in chemistry.

Stein, William H. (1911–80) American biochemist who worked with Moore to produce a new method of column chromatography by which they were able to separate and identify amino acids from a mix of material produced by the hydrolysis of proteins. Stein, Moore, and Anfinsen shared the 1972 Nobel Prize in chemistry.

Strohmeyer, Friedrich (1776–1835) German chemist who taught Bunsen and, in 1817, discovered the soft, bluish poisonous metal cadmium, now used in dry batteries, solders, and as a neutron absorber in nuclear reactors.

Sumner, James B. (1887–1955) American biochemist who crystallized the enzyme urease and proved that it was a protein. He then partly determined its mode of function and produced antibodies to it. He proceeded to investigate and purify a considerable range of enzymes active in human biochemistry. He shared the 1946 Nobel Prize in chemistry with Wendell Stanley and Northrop.

Svedberg, Theodor (1884–1971) Swedish chemist who studied colloidal chemistry, produced synthetic rubber, and developed the ultracentrifuge, which became an important instrument in chemical and biological research. For this, he was awarded the 1926 Nobel Prize in chemistry.

Sir Joseph Wilson Swan

Swan, Sir Joseph Wilson (1828–1914) English chemist and physicist who in 1860 invented the electric lamp 20 years before Edison. In 1864 he patented the carbon process for photographic printing; in 1871 he invented the dry-plate technique; and in 1879 he produced bromide paper. He was the first to produce a practicable artificial silk.

Synge, Richard L. M. (1914–94) British chemist who developed partition chromatography using two liquids that would not mix, one being held by the absorbing material (the stationary phase), the other being the moving phase that carries the samples. Synge used powdered cellulose or potato starch as the stationary phase in his columns. He was awarded the 1952 Nobel Prize in chemistry.

von Szent-Györgi, Albert (1893–1986) Hungarian-born American biochemist who found vitamin C in the adrenal gland and in paprika. He studied muscle tissue and action, isolated the two muscle contractile proteins actin and myosin, and investigated the role of the thymus gland. He showed how adenosine triphosphate (ATP) caused these proteins to contract, and he helped to elucidate the metabolic and energetic processes in the muscle cell. He was awarded the 1937 Nobel Prize in physiology or medicine.

Albert von Szent-Györgi

Takamine, Jokichi (1854–1922) Japanese-born American chemist who in 1901 isolated a substance for adrenal glands that was shown to be epinephrine (adrenaline). This was the first isolation of a pure hormone.

Tartaglia, Niccolo (1500–57) Italian mathematician who found a way of solving equations containing a cube of the unknown. His first book, *Nova scientia* (1537), dealt with ballistics, falling bodies, and projectiles, and showed that a firing angle of 45 degrees gave the maximum range for a gun. He also wrote a three-volume work entitled *Treatise on Numbers and Measurements* (1556–60).

Henry Taube

Taube, Henry (1915–94) Canadian-born American chemist who invented a method for studying the transfer of electrons during chemical reactions. He also showed that metal ions in solution form chemical bonds with water. He was awarded the 1983 Nobel Prize in chemistry.

Tennant, Smithson (1761–1815) English chemist who proved by burning a diamond that it was a form of carbon and, while studying platinum for commercial purposes, isolated two new elements, iridium and osmium.

Axel Hugo Theodor Theorell

Theorell, Axel Hugo Theodor (1903–82) Swedish biochemist who crystallized muscle hemoglobin, investigated enzymes such as peroxidases and dehydrogenases, invented an electrophoresis (electrical attraction) arrangement for separating proteins of different molecular weight, and introduced fluorescence spectrometry. His work earned him the Nobel Prize in medicine or physiology in 1955.

Arne Wilhelm Kaurin Tiselius

Tiselius, Arne Wilhelm Kaurin (1902–71) Swedish chemist who separated the blood proteins into albumins and alpha, beta, and gamma globulins, using an electrophoresis apparatus designed by himself. He showed that antibodies were gamma globulins, and he developed chromatographic methods for the separation and identification of amino acids. In 1948, he was awarded the Nobel Prize in chemistry.

Todd, Sir Alexander R., (Baron Todd of Trumpington) (1907–97) Scottish biochemist who worked on the chemistry of a range of vitamins and other natural products and who showed how the four bases (adenine, guanine, cytosine, and thymine) were attached to sugar and phosphate groups. This is how DNA is formed, and various combinations of these bases, taken three at a time, form the genetic code. For this work, Todd was awarded the 1957 Nobel Prize in chemistry.

Travers, Morris William (1872–1961) English chemist who with Ramsay discovered krypton, neon, and xenon (1894–1908). Travers isolated these gases from liquid air, a procedure that required very low temperatures, and he developed a way of liquefying hydrogen. He also established the properties of argon and helium and investigated the properties of gases at very high temperatures.

Trembley, Abraham (1710–84) Swiss naturalist who studied the grafting and regeneration of animal tissue but with only limited success.

Urbain, Georges (1872–1938) French chemist who specialized in the study of the rare earth elements. After enormous labor involving hundreds of thousands of fractional crystallizations, he discovered samarium, europium, gadolinium, terbium, holmium, lutetium, and hafnium.

Urey, Harold Clayton (1893–1981) American chemist who discovered deuterium (heavy hydrogen), the isotope of hydrogen that has a neutron and a proton in the nucleus and is thus twice the weight of common hydrogen, which has only a proton. Urey also worked with Miller on experiments to simulate a primitive atmosphere, thought to be similar to Earth's early atmosphere. Urey was awarded the Nobel Prize in chemistry in 1934.

Harold Clayton Urey

du Vigneaud, Vincent (1901–78) American biochemist whose principal contribution was in the field of amino acids. He showed how a series of these important protein constituents were synthesized in the body. He also achieved the laboratory synthesis of thiamine and penicillin, and the hormones oxytocin and vasopressin. He was awarded the 1955 Nobel Prize in chemistry.

Virtanen, Artturi I. (1895–1973) Finnish biochemist who discovered the chemical pathways by which bacteria in certain plant root nodules can achieve the fixation of nitrogen into compounds usable by plants. This work earned him the 1945 Nobel Prize in chemistry.

Johannes Diderik van der Waals

van der Waals, Johannes Diderik (1837–1923) Dutch physical chemist who, aware that real gases did not accurately conform to the simple gas law pV=RT (pressure x volume = temperature x the gas constant R), devised a more precise gas law equation that took account of the volume of the gas molecules and the attraction between them. He received the Nobel Prize in physics in 1910.

Otto Wallach

Walker, John E. (b. 1941) British chemist whose studies on the detailed structure of the enzyme ATP synthase (ATPase) confirmed Paul Boyer's account of the function of this important enzyme and earned him a share, with Boyer, of the 1997 Nobel Prize in chemistry.

Wallach, Otto (1847–1931) German chemist who studied volatile oils and, from this work, established and named the terpene class of compounds. He showed that the terpenes consisted of a variable number of five-carbon units, each of which was called an isoprene unit. He was awarded the 1910 Nobel Prize in chemistry.

James Dewey Watson

Watson, James Dewey (b. 1928) U.S. bird expert (ornithologist) who worked at the Cavendish Laboratory, Cambridge, England, with Crick and, in 1953, made the greatest biological discovery of the 20th century: the structure of DNA. Their joint paper in *Nature* is one of the most important scientific communications ever made, and it revolutionized genetics and molecular biology. Watson shared with Wilkins and Crick the Nobel Prize in physiology or medicine in 1962, and in 1988 became head of the Human Genome Project to sequence the whole of human DNA.

Werner, Alfred (1866–1919) Swiss chemist who was the first to point out that isomerism (same molecular formula but different molecular structure) occurred in inorganic as well as in organic chemistry. This was at first rejected by his associates but was later accepted and was highly influential on the progress of chemistry. He also advanced valency theory. For his work, he was awarded the 1913 Nobel Prize in chemistry.

Heinrich O. Wieland

Wieland, Heinrich O. (1877–1957) German chemist who studied the constitution of bile acids, organic radicals, nitrogen compounds, and contributed to the advancement of organic chemistry. He was awarded the 1927 Nobel Prize in chemistry.

Wilkins, Maurice Hugh Frederick (b. 1916) New Zealand-born British biophysicist who, with Crick, Watson, and Rosalind Franklin, worked to determine the molecular structure of DNA by X-ray crystallography. Wilkins shared the Nobel Prize in physiology or medicine with Crick and Watson in 1962.

Wilkinson, Sir Geoffrey (1921–96) British chemist whose main work was in the investigation of the organometallic compounds. His study of ferrocene showed that its molecule consisted of an atom of iron sandwiched between two five-sided rings of carbons and hydrogens. Thousands of substances with this kind of structure have since been synthesized and found to be of great chemical importance. He was awarded the 1973 Nobel Prize in chemistry with Ernst Fischer.

Willstatter, Richard M. (1872–1942) German chemist who carried out extensive investigations into the structure of chlorophyll and other flower pigments. He was awarded the Nobel Prize in chemistry in 1915.

Windaus, Adolf O. R. (1876–1959) German biochemist who studied the drug digitalis, established the structure of cholesterol, and researched vitamin D and some of the B vitamins. He was awarded the 1928 Nobel Prize in chemistry.

Adolf O. R. Windaus

Winkler, Clemens (1838–1904) German chemist who discovered germanium, one of the elements predicted by Mendeleyev on the basis of the periodic table. Germanium became well known when the physicists William Shockley, Walter Brattain, and John Bardeen used it to make a point contact rectifier and then the first transistor.

Wittig, Georg (1897–1987) German organic chemist who discovered that some organometallic reagents with both negative and positive charges could react readily with aldehydes and ketones to simplify the synthesis of a useful range of organic compounds. He was awarded the 1979 Nobel Prize in chemistry.

Wöhler, Friedrich (1800–82) German chemist whose synthesis of urea from ammonium cyanate in 1828 was the first time a compound of organic origin had been prepared from inorganic material. This achievement revolutionized organic chemistry and showed that living organisms were not fundamentally different in structure from nonliving matter. Wöhler also isolated aluminum in 1827 and beryllium in 1828 and discovered calcium carbide, from which he obtained acetylene.

William H. Wollaston

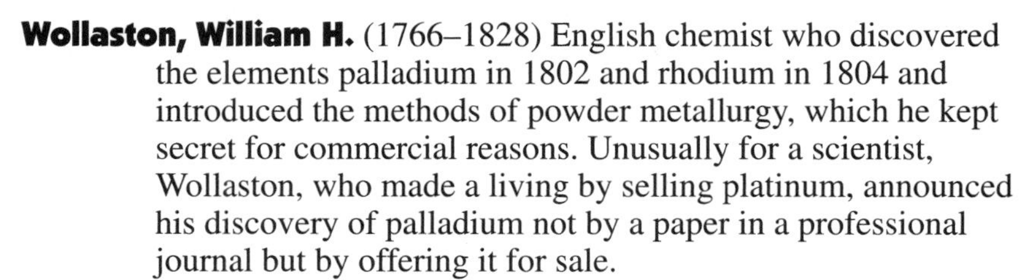

Wollaston, William H. (1766–1828) English chemist who discovered the elements palladium in 1802 and rhodium in 1804 and introduced the methods of powder metallurgy, which he kept secret for commercial reasons. Unusually for a scientist, Wollaston, who made a living by selling platinum, announced his discovery of palladium not by a paper in a professional journal but by offering it for sale.

Woodward, Robert B. (1917–79) American chemist who is generally regarded as the greatest synthetic chemist of all time. He achieved the synthesis of a long succession of biochemical substances, including cholesterol, cortisone, chlorophyll, reserpine, strychnine, quinine, the antibiotics cephalosporin and tetracycline, and cyanocobalamin (vitamin B_{12}). For this work, he was awarded the 1965 Nobel Prize in chemistry.

Young, James (1822–93) Scottish chemist and industrialist who showed that low-temperature distillation of shale could yield substantial commercial quantities of paraffin oil and solid paraffin wax. This led to a shale oil industry in Scotland and a fortune for Young, who sold his shale oil business in 1866 for $2 million.

Karl Ziegler

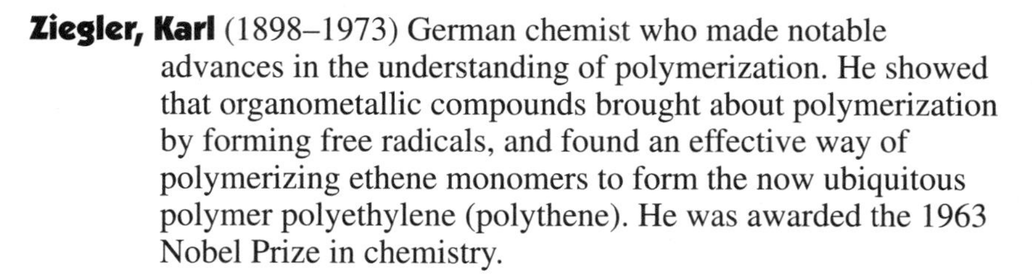

Ziegler, Karl (1898–1973) German chemist who made notable advances in the understanding of polymerization. He showed that organometallic compounds brought about polymerization by forming free radicals, and found an effective way of polymerizing ethene monomers to form the now ubiquitous polymer polyethylene (polythene). He was awarded the 1963 Nobel Prize in chemistry.

Zsigmondy, Richard A. (1865–1929) Austrian chemist who invented the dark ground illumination method of microscopy and showed that color changes in colloidal solutions were due to particle aggregation. He made many advances in colloidal chemistry for which he was awarded the 1925 Nobel Prize in chemistry.

SECTION THREE
CHRONOLOGY

The elements carbon, copper, gold, iron, lead, mercury, silver, sulfur, and tin are all used and known in prehistoric times, although their status as elements is not realized

7000–6000 BCE • Lead and copper production developed in Anatolia, Turkey

3500 BCE • Copper and bronze production spreads throughout the Middle East

2700 BCE • Fabrication of iron objects (from meteoric iron) is evidenced

2500 BCE • Standardized weights used in Sumeria. Silver sheet-metal working is carried out

c. 2500 BCE • Glass making occurs in Mesopotamia

2000-1000 BCE • Hittites develop iron technology

c. 1550 BCE • Glass first made in Egypt

950-500 BCE • First Iron Age in Europe. Extraction and working of iron spread gradually across the continent

c. 600 BCE • Tin mining in Cornwall. In China, Lao Tzu explains his philosophy, known as Taosim, in the *Tao Te Ching* (The way of life). The universe is seen in terms of opposites, "yang," the male, positive, hot, and light principle, and "yin," the female, negative, cool, and dark principle. The five elements, earth, water, fire, metal, and wood, are believed to be generated by the struggle between these opposing forces. The Chinese produce gunpowder and are thought to have been able to produce nitric acid. In Greece, the theory that all substances are generated from a single primary matter, a featureless substance, is proposed

580 BCE • Early theory of matter proposed by Greek philosopher Thales, suggesting that all things are made from forms of water

c. 569 BCE • Evidence of the use of bellows for metallurgical workings, Anacharsis, Sythia

c. 560 BCE • Matter explained in terms of cold, heat, dryness, and wetness by Greek philosopher Anaximander

c. 530 BCE • Theory that all matter is made of air proposed by Greek philosopher Anaximines. According to the theory air is condensed under varying conditions to make all known liquids and solids

c. 500 BCE • Chinese use bronze and copper nickel alloys. Steel produced in India

480–471 BCE • Greek philosopher Anaxagoras suggests materials are made up of large numbers of "seeds," particles that determine qualities

450 BCE • Greek philosopher Empedocles (d. 430 BCE) introduces the four-element theory of matter (fire, air, water, earth). According to this theory, water (moist and cold), air (moist and hot), fire (dry and hot), and earth (dry and cold) combine in various ways and proportions to generate all of the materials found in the universe

c. 445 BCE • Atomic theory of matter introduced by Greek philosopher Leucippus

430–421 BCE • Concept of atoms is expanded by Greek philosopher Democritus

350–341 BCE • Greek philosopher Aristotle defines chemical elements as constituents of bodies that cannot be decomposed into other constituents

310–301 BCE • Greek philosopher Epicurus founds philosophical school based on a theory of atoms, particles so small that they cannot be subdivided further

150 BCE • Use of bellows in metallurgic furnaces

60 BCE • Roman poet and philosopher Lucretius writes on Greek atomic theory

c. 560 BCE
Anaximander explains his concept of matter.

105 CE • Paper made from vegetable fibers by Chinese court official Ts'ai Lun

c. 185 • Earliest known work on alchemy, forerunner of chemistry, compiled in Egypt

200 • Practice of casting iron established in China

400-409 • Term *chemistry* used for first time by Alexandrian scholars to describe the activity of changing matter

c. 185
Drawing of stills and furnaces, as used in alchemy.

789 • Standard units of weight and measure introduced by Charlemagne, Frankish king, later emperor of the West

880 • Persian physician and alchemist Rhazes produces the text *Secret of Secrets*. This work describes various laboratory procedures basic to the study of chemistry—purification, separation, mixing, removal of water, and solidification—and describes many items of laboratory equipment recognizable today. It also classifies substances into animal, vegetable, mineral, and derivative. Substances are further classified as metals, vitriols, boraxes, salts, and stones. He prepares sal ammoniac by distilling hair with salt and urine

13th century
China originates use of gunpowder-propelled rockets.

1140 • Iron industry established in Europe

c. 1250 • Description and manufacture of element arsenic by German scientist Albertus Magnus

c. 1300 • Alum discovered at Roccha in Syria. Sulfuric acid described

c. 1340 • Blast furnaces being used in Europe

1500 • Paracelsus (Philippus Aureolus Theophrastus Bombast von Hohenheim) develops new study of iatrochemistry, use of chemistry in medicine. He introduces the doctrine of the *tria prima*, that medical substances are made up of the four Aristotelian elements

1540 • Ether discovered by German scientist Valerius Cordus

1592 • Early thermometer invented by Italian scientist Galileo Galilei

1597 • Andreas Libavius (1540-1616) publishes *Alchemia*, said to be the first chemistry textbook. It includes classification and standardization of laboratory techniques and substances used in the study of chemistry

c. 1600 • Discovery of element antimony. Belgian scientist Joannes Baptista van Helmont proposes two elements, air (a physical medium) and water (the material from which all substances are formed). He says that fire is the agent of change. He conducts an experiment to grow a willow tree in controlled environment, and deduces that any weight gain in the plant is solely caused by water. He also suggests the existence of a universal solvent (alcahest) that others try to isolate without success. Defines gas as "This spirit, hitherto unknown, which can neither be retained in vessels or reduced to a visible body . . . I call by the new name gas"

1637 • Philosophical study of scientific method published by French philosopher René Descartes, who also publishes a theory of refraction

1649 • Atomic theory of matter revived by French philosopher Pierre Gassendi. Publication of his book *Syntagma Philosophicum* in which he introduces the term *molecule* to indicate the smallest unit of a substance capable of an independent existence

1654 • Invention of the air pump by German phycisist Otto von Guericke. He demonstrates how air can be pumped out of a copper globe to leave a vacuum. He finds that the atmosphere exerts a tremendous compressing force on the globe (demonstrated in 1657 in the Magdeburg experiment where two metal hemispheres placed together and then evacuated with a pump could not be pulled apart by a team of 16 horses)

1597
Alchemical symbols are used in Islam and the west. The sun was linked with gold, the moon with silver, in a drawing of the tree of universal matter.

1657
Otto von Guericke demonstrates the force of atmosphere on a globe.

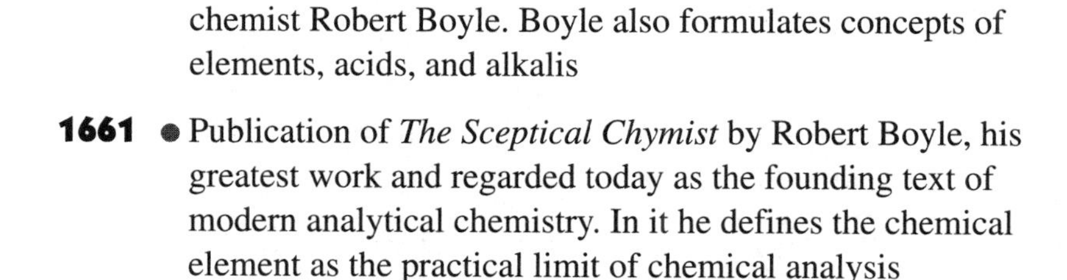

1660 • Publication of *New Experiments Physico-Mechanical touching the Spring of the Air and its Effects* by Irish chemist Robert Boyle. Boyle also formulates concepts of elements, acids, and alkalis

1661 • Publication of *The Sceptical Chymist* by Robert Boyle, his greatest work and regarded today as the founding text of modern analytical chemistry. In it he defines the chemical element as the practical limit of chemical analysis

1662 • Robert Boyle formulates Boyle's law, which states that the pressure and volume of a gas are inversely proportionate

1669 • Element phosphorus discovered by German chemist Hennig Brand. Johann Joaquim Becher publishes *Physica Subterranea*, the first attempt to relate observations and concepts in physics and chemistry

1670 • Formic acid discovered by English biologist John Ray

1676 • English chemist Robert Hooke formulates Hooke's law relating to elastic bodies

1692 • Newton writes *De natura acidorium*. In this and other writings he suggests that there are exceedingly strong attractive powers between the particles of bodies that extend for a short range and vary in strength between chemical species. This leads to the idea of "elective affinities" concerning the replacement of one metal by another in acid solutions. He makes a list giving the replacement order of six common metals in nitric acid

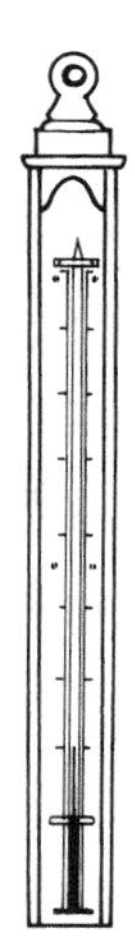

1714
Gabriel Fahrenheit's original thermometer.

1702 • Boracic or boric acid discovered by German chemist Wilhelm Homberg

1714 • First practical mercury thermometer invented by German physicist Gabriel Fahrenheit, who also devises the Fahrenheit scale

1718 • Etienne Geoffroy (1672–1731) produces the first table of

affinities (following Newton's work on elective affinities). Georg Stahl (1660–1734) proposes the phlogiston theory. Phlogiston is described as a substance present in all materials that are able to be burned and that is released on burning. Stahl also believes that vitriolic acid (sulfuric acid) is the universal solvent

1723 • First-known treatise on crystallography published by M. A. Capeller

1727 • English chemist Stephen Hales (1677–1761) establishes that air takes part in chemical reactions

1729 • Electricity found to travel through conductors by English physicist Stephen Gray

1730 • Alcohol thermometer developed by French physicist René Antoine Ferchault de Réaumur

1735 • Discovery of element platinum in South America by Spanish scientist Antonio de Ulloa. Discovery of element cobalt by Swedish chemist Georg Brandt (1694–1768)

1742 • Celsius (centigrade) temperature scale devised by Swedish astronomer Anders Celsius, although with boiling point at zero and freezing point at 100 degrees. The technique of galvanizing is invented by French metallurgist Paul Jacques Malouin

1744 • Nature of heat described as a form of motion by Russian scientist Mikhail Lomonosov (1711–65)

1745 • Invention of Leyden jar (for the storage of static electricity), independently, by Dutch physicist Pieter van Musschenbroek and German physicist Ewald Georg von Kleist. Ancestor of the modern capacitor

1746 • English chemist John Roebuck of Birmingham designs a large-scale process for manufacturing sulfuric acid using large lead-lined wooden boxes (the lead-chamber method)

1742
Swedish stamp (1982) with a portrait of Anders Celsius and a representation of his temperature scale.

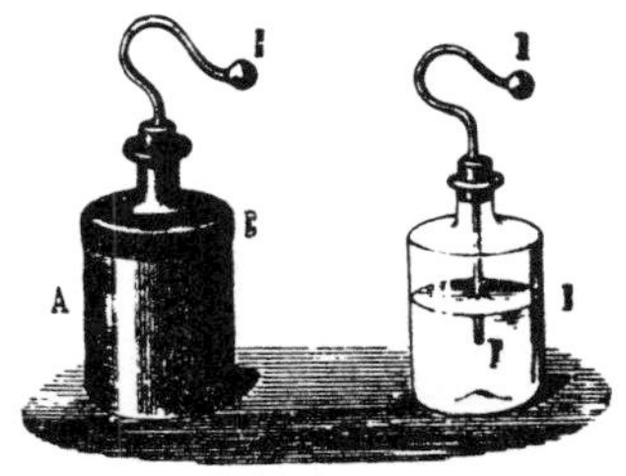

1745
Leyden jar invented.

1747 • Sugar produced from raw beetroot by German scientist Andreas Marggraf

1748 • Osmotic pressure (relating to solutions) discovered by French physicist Jean Nollet

c. 1750 • French chemist Guillaume-François Rouelle (1703–70) proposes a theory classifying salts by their crystalline shape and by the acids and bases from which they are produced. Uses the term *water of crystallization*

1751 • Element nickel discovered by Swedish mineralogist Axel F. Cronstedt. He also demonstrates its magnetic properties

1753 • The element bismuth is discovered

1756
Joseph Black discovered carbon dioxide.

1756 • Scottish chemist Joseph Black (1728–99) shows that carbon dioxide (carbonic acid gas) is different from ordinary air. He demonstrates that magnesium carbonate contains a gas (carbon dioxide) that is different from atmospheric air; it turns lime water milky and does not support burning

1761 • In his study of calorimetry, Joseph Black establishes the concepts of latent heat and specific heat capacity

1762 • Iron smelting with coal rather than charcoal developed by John Roebuck, at the Carron Ironworks, Scotland

1766 • Element hydrogen discovered by English chemist Henry Cavendish

1770 • Tartaric acid discovered by Swedish chemist Carl Wilhelm Scheele (1742–86)

1771 • Element fluorine discovered by Carl Wilhelm Scheele. Picric acid discovered by British chemist Peter Woulfe

1772 • Element nitrogen discovered independently by Scottish chemist Daniel Rutherford, English chemist Joseph Priestley, and two other chemists. French scientist Jean Rome de Lisle describes process of crystallization. Laughing gas (nitrous oxide) discovered by Joseph Priestley

1773 • French chemist Antoine Lavoisier publishes *The Analytic Spirit*, suggesting the existence of three distinct states of matter – solid, fluid, and the state of expansion of vapors. The same body can exist in each of the forms, depending on the quantity of the matter of fire combined with it

1774 • Discovery of element manganese credited to Swedish mineralogist Johan Gottlieb Gahn. Composition of the atmosphere measured by Joseph Priestley. Elements oxygen and chlorine separated by Carl Wilhelm Scheele. Chlorine used for bleaching in Sweden

1774
Joseph Priestley discovers hydrochloric and sulfuric acids.

1778 • Discovery of element molybdenum by Carl Wilhelm Scheele

1779 • Glycerin discovered by Carl Wilhelm Scheele

1780 • Calorimeter developed by French chemists Pierre Simon de Laplace and Antoine Lavoisier

1781 • English chemist Joseph Priestley sparks "inflammable air" (hydrogen) with air using an electrostatic machine. He discovers that water is produced. The element tungsten is recognized by Carl Wilhelm Scheele in the ore subsequently named after him, scheelite

1782 • Artificial ice made chemically by English chemist Walker. Discovery of element tellurium by Austrian mineralogist Franz Joseph Müller (1740–1825). Guyton, Lavoisier, Berthollet, and Fourcroy publish *Méthode de nomenclature chimique*. Substances have one fixed name: the name reflects composition and names are chosen from Greek and Latin roots

1785 • Lavoisier attacks phlogiston theory as being a "veritable Proteus that changes its form every instant!" (in order to explain various apparently contradictory phenomena)

1787 • Charles' law (connecting the expansion of gas with its rise in temperature) propounded by French physicist Jacques Charles

c. 1788 • Lavoisier shows water consists of a combination of hydrogen and oxygen

1789 • French chemist Nicholas Leblanc devises a process for the manufacture of soda. Publication of Lavoisier's *Traite élémentaire de chimie* (An elementary treatise on chemistry) in two volumes. In this he defines a chemical element as a substance that cannot be analyzed by chemical means. He makes a list of 33 elements and divides them into five classes. A compound of uranium discovered by German chemist Martin Heinrich Klaproth

1790 • Introduction of Leblanc's process for soda manufacture

1791 • Discovery of element titanium by English chemist William Gregor

1792 • German chemist Jeremias Richter studies the mass ratios of substances combining together, which leads to the Law of Reciprocal Proportions and to the subject of stoichiometry (a branch of chemistry concerned with the proportions in which elements are combined in compounds)

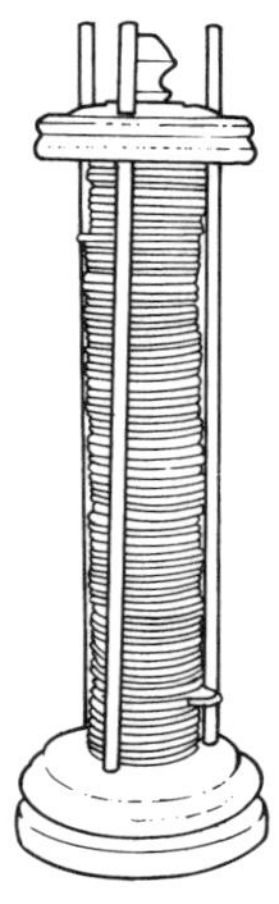

1800
Alessandro Volta uses "piles" of alternating zinc and silver disks in the first battery.

1797 • Discovery of element chromium by French chemist Louis-Nicolas Vauquelin

1798 • Discovery of element beryllium by Louis-Nicolas Vauquelin

1790–1799 • Metric system devised by a committee of seven Frenchmen to regulate scientific measurements

1799 • Foundation of Royal Institution in London by Benjamin Thompson (also known as Count Rumford). The aim of the institution is to publicize ways in which science can be used to improve the quality of life

1800 • English chemists William Nicholson and Anthony Carlisle use electricity to produce chemical change. Electric cell invented by Italian physicist Alessandro Volta. The electric cell, "voltaic pile," is made from alternating zinc and silver disks and is regarded as the first battery. This early battery can decompose water into hydrogen and oxygen at the different poles of the battery. The element zinc is discovered

1801 • Element niobium discovered by English chemist Charles Hatchett. English chemist John Dalton formulates law on gas pressure (partial pressures in a mixture of gases) in his paper "New theory of the constitution of mixed aeriform fluids and particularly of the atmosphere." Henry's law (that the amount of gas absorbed by a liquid varies directly with the pressure) formulated by English chemist and physicist William Henry. Element vanadium discovered by Spanish mineralogist Andrès del Rio. He names it erythronium; it is later named vanadium by Nils Sefström, Swedish chemist, who discovers it independently in 1831

1802 • Discovery of element tantalum by Swedish chemist Anders G. Ekeberg. English chemist John Dalton makes atomic weight tables. Electromagnetism discovered by Danish physicist Hans Oersted

1803 • Claude Berthollet suggests that mass (concentration), temperature, and pressure have an effect on chemical reactions. Discovery of elements palladium and rhodium by English chemist William H. Wollaston. Discovery of element cerium by Swedish chemist Jons Jakob Berzelius, Swedish mineralogist Wilhelm Hisinger, and German chemist Martin H. Klaproth. The theory that matter is made up of atoms is proposed, not for the first time in the history of science, by English chemist John Dalton. Dalton publishes his "table of the relative weights of the ultimate particles of gaseous and other bodies." This is the first table of atomic weights. The elements iridium and osmium discovered by English chemist Smithson Tennant

1804 • Studying electrolysis, both Davy and Berzelius (with William Hisinger) conclude that "combustible bodies and bases" are released at the negative pole; oxygen and oxidized bodies are released at the positive pole

1805 • Composition of water (hydrogen and oxygen) established by French chemist Joseph Gay-Lussac

1807 • Discovery of elements sodium and potassium by English chemist Humphry Davy. English physicist Thomas Young is the first to use the word "energy" with a meaning close to its modern sense

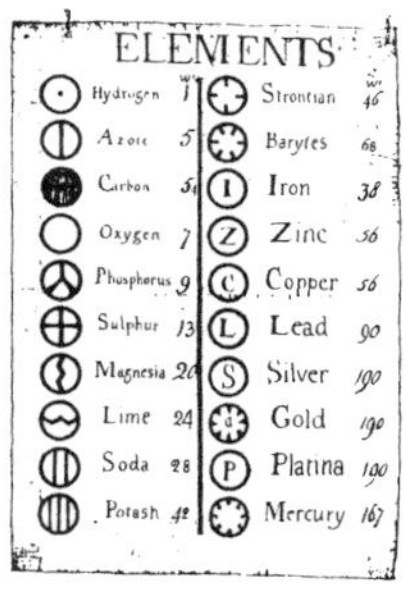

1808
John Dalton's symbols for elements in A New System of Chemical Philosophy.

1808 • Elements barium, magnesium, and calcium discovered by English chemist Humphry Davy. Gay-Lussac's law, "Gases combine amongst themselves in very simple proportions," proposed by Joseph Gay-Lussac. Publication of John Dalton's *A New System of Chemical Philosophy*, in which Dalton suggests that matter is composed of a myriad of homogeneous atoms and that each element's atoms differ slightly in mass. He also states that elements compound together in fixed proportions by weight, based on the findings of the new stoichiometric approach to chemistry. Dalton regards chemical reactions as the reordering of atoms into new groups (or molecules). Dalton has four basic assumptions. All matter is composed of solid, indivisible atoms. Atoms retain their identity in all chemical reactions; they are indestructible. There are as many types of atoms as there are elements. Atoms of different elements combine to make compound atoms (a process he calls chemical synthesis) in certain fixed ratios. He determines atomic weights of elements relative to element hydrogen (the lightest known). Element boron is isolated by Humphry Davy. Davy also isolates element strontium using electrolysis. It is named after Strontian, a Scottish village where it is first found

1808
Dalton explains atomic theory.

1810 • English chemist Humphry Davy proves chlorine to be an element and names it

1811 • Avogadro's hypothesis, "Equal volumes of gas contain the

same number of molecules" (defined as stable multi-atomed particles), proposed by Italian chemist Amedeo Avogadro. Avogadro also first uses the term *molecule*. Discovery of element iodine by French chemist Bernard Courtois while studying the liquor obtained in leaching the ashes of burnt kelp. Chemical symbols devised by Jons Jakob Berzelius. As a result of his experiments on the electrolysis of various solutions, Berzelius devises Latin classification of substances based on electrochemical phenomena. Substances are divided into two types, electropositive and electronegative. He defines electropositive substances as those attracted to the negative pole and introduces some basic terms in the study of electromagnetic phenomena

1815 • English chemist William Prout (1785–1850) calculates the specific gravities of elements using air as the standard and then comparing the result with that of hydrogen (assumed to have specific gravity of one). His calculations produce whole numbers, and he suggests that hydrogen could be the basis of all matter ("Prout's hypothesis")

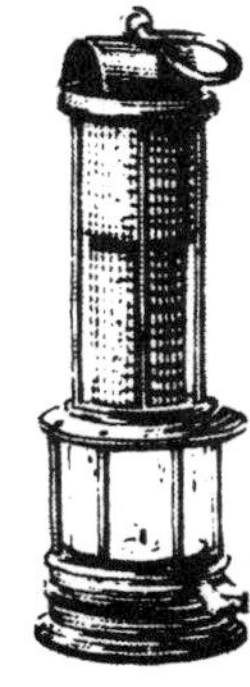

1815
Humphry Davy invents safety lamp for use in mines.

1817 • Discovery of element lithium by Swedish chemist Johan August Arfwedson. Discovery of element cadmium by German chemist Friedrich Strohmeyer. Discovery of element selenium by Swedish chemist Jons Jakob Berzelius

1818 • Discovery of the phenomenon of isomorphism in inorganic crystals by German chemist Eilhard Mitscherlich

1819 • Naphthalene discovered by British chemist A. Garden

1822 • Mohs' scale for mineral hardness proposed by German metallurgist Friedrich Mohs

1823 • Liquefaction of gases achieved by English chemist Humphry Davy (British). Discovery of element silicon by Swedish chemist Jons Jakob Berzelius

1825 • Isolation of benzene (benzol) by English scientist Michael Faraday from compressed oil gas

1826 • Element bromine is discovered by French chemist Antoine-Jérôme Balard. Aniline discovered by German chemist Otto Unverdorben by the distillation of indigo. Limelight (when heated, calcium oxide or lime becomes incandescent) is invented by Scottish scientist Thomas Drummond

1827 • Aluminum isolated by German chemist Friedrich Wöhler. The invention of the friction match (Lucifer matches) by English scientist John Walker

1828 • Discovery of element thorium by Swedish chemist Jons Jakob Berzelius. Friedrich Wöhler synthesizes urea from ammonium cyanate. This is the first time a compound of organic origin has been prepared from inorganic material. The value of using a hot-blast (precursor of the blast furnace, in iron smelting) is discovered by Scottish metallurgist James Neilson. The element yttrium (named after a village in Sweden) is discovered

1829 • Döbereiner's triads: German chemist Johann Döbereiner shows the existence of groups of three chemically similar elements, or triads. The atomic weight of the central element in each triad is the arithmetic mean of the other two. Graham's law relating to the diffusion rate of gases is formulated by Scottish scientist Thomas Graham

Cl
|
H – C – Cl
|
Cl

1831
Structure of chloroform discovered by Soubeiran and Liebig.

1830 • The phenomenon of isomerism is identified by Berzelius to explain the lack of compositional difference between racemic and tartaric acids. Discovery of paraffin by German chemist Baron Karl von Reichenbach

1831 • Discovery of electromagnetic induction by Michael Faraday. Chloroform discovered simultaneously by French chemist Eugène Soubeiran and German chemist Baron Justus Liebig

c. 1833 • Creosote discovered by Baron Karl von Reichenbach. Terms *electrode*, *cathode*, *anode*, *ion*, *cation*, *anion*, *electrolyte*, and *electrolysis* coined by English physicists

Michael Faraday and William Whewell. Laws of electrolysis proposed by Michael Faraday. The first enzyme is discovered (diastase from barley) by French chemist Anselme Payen

1834 • Carbolic acid discovered in coal tar by German scientist Friedlieb Ferdinand Runge. He pioneers the field of chromatography

1836 • The first animal enzyme (pepsin) is discovered by German scientist Theodor Schwann (1810–82). The fuel cell invented by Welsh electrochemist Sir William Robert Grove

1837 • Justus Liebig introduces the convention of using subscripts to denote numbers of atoms of an element in a compound (for example H_2O). Electroplating invented by British metallurgist Thomas Spencer

1839 • Element lanthanum discovered by Swedish chemist Carl Gustaf Mosander. The effects of heating rubber with sulfur (vulcanization) are discovered by American inventor Charles Goodyear. Ozone discovered by German scientist Christian Friedrich Schönbein

1841 • French scientist Auguste Laurent isolates phenol in his study of coal tar derivatives. Allotropy in carbon discovered by Jons Jakob Berzelius

1842 • First use of ether as an anesthetic by American physician Crawford Long

1842–45 • French chemist Charles Gerhardt formulates the idea of homologous series

1843 • Process of manufacturing superphosphate patented by English agricultural scientist Sir John Lawes. Identification of elements erbium and terbium by Swedish chemist Carl Gustaf Mosander

1848
Louis Pasteur discovers anerobic organisms.

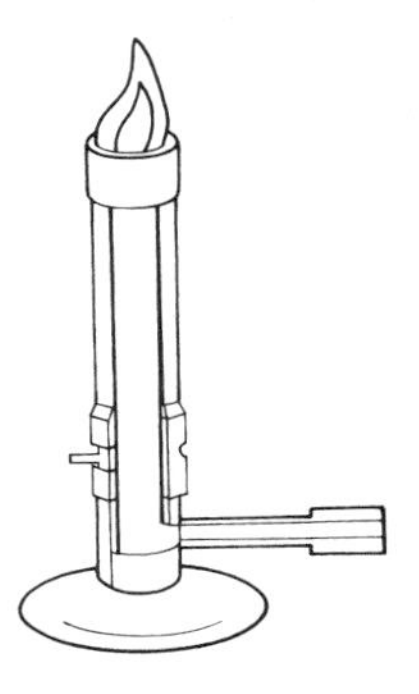

1850
Robert Bunsen's laboratory burner.

1844 • Discovery of the pure element ruthenium by Russian chemist Karl K. Klaus

1845 • Establishment of the Royal College of Chemistry

c. 1847 • Cathode rays discovered by German mathematician and physicist Julius Plücker. First law of thermodynamics stated by German physicist Hermann Ludwig Ferdinand von Helmholtz. Nitroglycerin discovered by Italian chemist Ascanio Sobrero

1848 • Foundation of the American Association for the Advancement of Science. Foundation of stereochemistry (the study of the spatial arrangement of atoms in molecules and the effect of these arrangements on chemical properties) by French scientist Louis Pasteur. Chloroform first used as an anesthetic by Scottish physician Sir James Simpson during experiments on himself. Ethane discovered by English chemist Sir Edward Frankland and German chemist Adolph Kolbe. Concept of absolute zero formulated by Scottish physicist William Thomson, later Baron Kelvin

1849 • Introduction of fractional distillation by Charles Mansfield. Term *thermodynamics* introduced by Scottish physicist William Thomson, Baron Kelvin

1850 • Second law of thermodynamics formulated by German physicist Rudolf Julius Emmanuel Clausius. Bunsen burner invented by German chemist Robert Bunsen. Regelation of ice discovered by Michael Faraday

1852 • Term *fluorescence* coined by Irish physicist George Stokes

1856 • Aniline dye, a brilliant purple dye, is invented by British chemist Sir William Perkin. The dye is marketed as "aniline purple" or "mauveine." Regenerative furnace designed by Anglo-German metallurgist Sir Charles William Siemens

1859 • Spectroscope (an instrument for recording a spectrum of electromagnetic radiation by dispersal) developed by Robert Bunsen and G. R. Kirchhoff. This enables the identification of new elements. Kinetic theory of gases developed by Scottish physicist James Clerk Maxwell

1860 • English metallurgist Sir Henry Bessemer invents the Bessemer converter for steel manufacture. Element cesium discovered by German chemists Robert Bunsen and Gustav Kirchhoff. Statistics for analyzing the behavior of molecules in a gas independently developed by Scottish physicist James Clerk Maxwell and Austrian physicist Ludwig Eduard Boltzmann. Magnesium light invented by Robert Bunsen

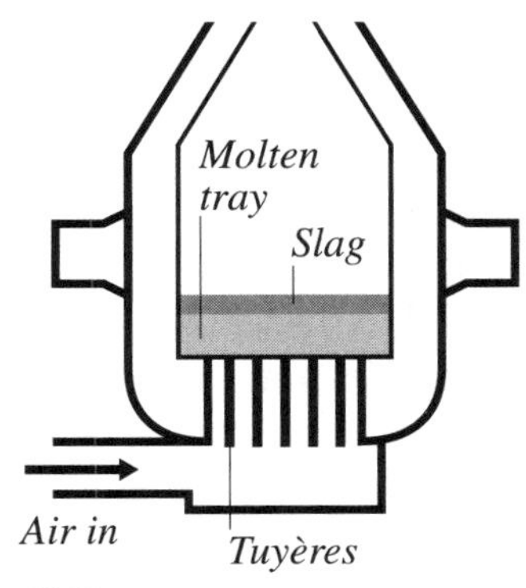

1860
Bessemer converter converts molten iron into steel.

1861 • Belgian chemist Ernst Solvay designs the Solvay tower for the production of soda. Carbon dioxide is forced up a tower down which flows ammoniated salt. The process uses brine and ammonia as raw materials. Element rubidium discovered by German chemists Robert Bunsen and Gustav Kirchhoff. Element thallium discovered by English physicist William Crookes

1862 • The first plastic (Parkesine) is prepared by British scientist Alexander Parkes. Nitrocellulose is mixed with wood naphtha (a mixture of methanol, acetone, acetic acid, and methyl acetate formed during the distillation of wood) to produce a malleable solid. It is marketed, with little success, as a sculpting material. Calcium carbide, from which he later obtains acetylene, first made by German chemist Friedrich Wöhler

1863 • Discovery of element indium by German mineralogists Ferdinand Reich and Theodor Richter. TNT (trinitrotoluene) discovered by German chemist J. Wilbrand. Production of amidoazobenzene (aniline yellow), the first yellow azo dye

1864 • English chemist Sir John Newlands arranges elements in order of ascending atomic weight and discovers that "the difference between the number of the lowest member of a group and that immediately above it is seven; in other words, the eighth element starting from a given one is a kind of repetition of the first, like the eight notes of an octave of music." He calls this the Law of Octaves. He is the first to assign a number to an element, leaving spaces for elements that are as yet undiscovered. Open-hearth steel production invented by German-born British electrical engineer Sir William Siemens and Pierre Émile Martin, a Frenchman

1865
Joseph Lister's carbolic acid sprayer.

1865 • Antiseptic surgery (using phenol) pioneered by English scientist Joseph Lister. German scientist Friedrich Kekulé (1829–96) suggests that benzene consists of a cyclic arrangement of six tetravalent carbon atoms with alternating single and double bonds between the carbon atoms

1867
The Nobel Prize struck in gold awarded for physics and for chemistry in memory of Alfred Nobel.

1867 • Dynamite invented by Swedish inventor, manufacturer, and philanthropist Alfred Nobel. This stable and safe explosive for industrial use makes him a fortune

1868 • German organic chemist Adolf von Baeyer demonstrates that complex organic molecules can be split into simpler compounds using heat and a zinc catalyst. Discovery of element helium in the solar spectrum by French astronomer Pierre-Jules-César Janssen and English astronomer Joseph Norman Lockyer. Zinc-carbon battery invented by French engineer Georges Leclanché. Coal tar perfumes invented by British chemist Sir William Perkin. Russian scientist Vladimir Markovnikov (1838–1904) discovers that in the hydrohalogenation of unsymmetrical unsaturated compounds, addition of the hydrogen takes place on the carbon with the most attached hydrogen, the halide added to the carbon with the least number of hydrogen atoms

1869 • Meyer's atomic volume curve: German chemist Julius Meyer plots the atomic volume (atomic weight divided by density) of each element against its atomic weight and shows that elements whose chemical properties are similar appear in similar positions on the waves of the curve. First periodic table for chemical elements published by Russian chemist Dmitri Mendeleyev. He writes that "The properties of the elements are in periodic dependence upon their atomic weight." His table shows how the elements are related to each other and how increasing atomic weight affects their chemical reactivities. Gaps in the table indicate elements yet to be discovered and predicts their properties. Synthetic alizarin developed for manufacture by William Perkin from the madder plant

1875 • Element gallium discovered by French chemist Paul-Émile Lecoq de Boisbaudran. The German industrial chemist Rudolph Messel perfects the "contact" process for sulfuric acid manufacture

1876 • German physicist Eugen Goldstein discovers cathode rays

1877 • French chemist Charles Friedel and American chemist James Mason Crafts discover that an aluminum chloride catalyst transforms organic chlorides into hydrocarbons and acid halides into ketones. This becomes known as the Friedel-Crafts reaction and is important in chemical synthesis

1878 • Element holmium discovered by Swiss scientists J. L. Soret and Marc Delafontaine

1879 • Element scandium discovered by Swedish chemist Lars Nilson. Element samarium discovered by French chemist Paul-Émile Lecoq de Boisbaudran. Element thulium discovered by Swedish chemist Per Cleve

1880 • Element gadolinium discovered by Swiss chemist Jean-Charles de Marignac. Piezoelectricity discovered by French chemist Pierre Curie

1881 • Concept of electromagnetic mass introduced by English physicist Sir Joseph John Thomson

1883 • Artificial silk made by British inventor Sir Joseph Swan

1881
Portrait of Sir Joseph John Thomson who devised the concept of electromagnetic mass.

1884 • Swedish physical chemist Svante Arrhenius proposes in his doctoral thesis the theory of electrolytic dissociation or ionization (acids, bases, and salts in solution dissociate into ions)

1885 • Discovery of elements neodymium and praseodymium by Austrian chemist Carl Auer. Peter Laar describes and names the phenomenon of tautomerism

1886 • Element germanium discovered by German chemist Clemens Winkler. Element dysprosium discovered by French chemist Paul-Émile Lecoq de Boisbaudran. Extraction of aluminum by electrolysis of aluminum oxide independently achieved by American scientist Charles Hall and French metallurgist Paul Héroult. Isolation of element fluorine by French chemist Henri Moissan. Melinite invented by French chemist Eugène Turpin. Saccharine invented by Americans C. Fahlberg and Ira Remsen

1887 • American chemist Arthur Michael discovers the "Michael condensation" reaction – the transformation of an unsaturated compound into a saturated compound with an additional carbon atom. German physical chemist Friedrich Wilhelm Ostwald formulates "Ostwald's dilution law." Photoelectric effect discovered by German physicist Heinrich Rudolf Hertz

1889 • Sir Frederick Augustus Abel develops cordite, an explosive that is safe to handle

1892 • American chemist Hamilton Young Castner develops an electrolytic cell for the formation of sodium hydroxide. This is later improved by Austrian engineer Karl Kellner, who devises a mercury cathode. German organic chemist Adolf von Baeyer introduces *cis-trans* terminology to the study of isomers. British chemical technologist Charles Cross invents the viscose rayon production process

1894 • Element argon discovered by Scottish chemist William Ramsay and English scientist Lord Rayleigh (John William Strutt). Liquefaction of oxygen discovered by Scottish scientist Sir James Dewar

1895 • Cloud chamber, an instrument for studying high energy particles by detecting their tracks through an enclosed medium, developed by Scottish physicist Charles Thomson Rees Wilson. Scottish chemist William Ramsay is first to discover element helium on Earth from the mineral clevite. X rays discovered by German physicist Wilhelm Konrad Roentgen

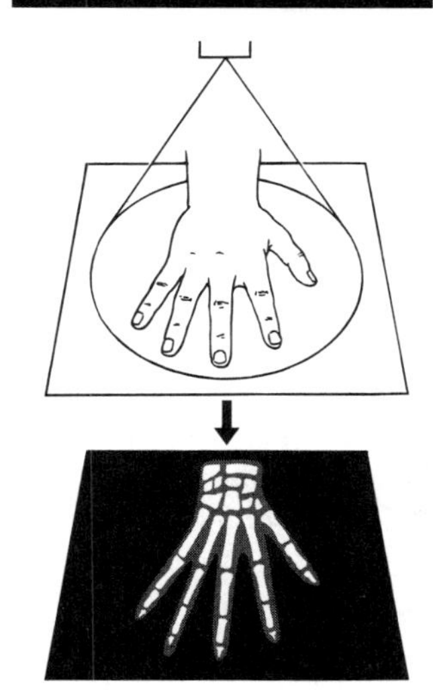

1895
Wilhelm Roentgen discovers X-rays.

1896 • Zeeman effect discovered by Dutch physicist Pieter Zeeman. Link between global temperature and atmospheric carbon dioxide made by Swedish physical chemist Svante Arrhenius. Radioactivity discovered by French physicist Henri Becquerel. Element europium discovered by French chemist Eugène-Anatole Demarçay

1897 • Castner-Kellner process for sodium hydroxide manufacture introduced. Electron discovered and its mass calculated by English physicist Sir Joseph John Thomson

1898 • Elements radium and polonium discovered by French chemist Pierre Curie and French physicist Marie Curie. Term *radioactivity* coined by Marie Curie after her study of these elements. Discovery of elements krypton, neon, and xenon by Scottish chemist William Ramsay and English chemist Morris William Travers

1898
Marie Curie with her equipment.

1899 • Alpha and beta rays (later known as alpha and beta particles), two types of radioactivity, distinguished by British nuclear physicist Sir Ernest Rutherford (later Lord Rutherford of Nelson, New Zealand). Discovery of element actinium by French chemist André-Louis Debierne. "Lock and key" hypothesis for enzymes made by German organic chemist Emil Fischer

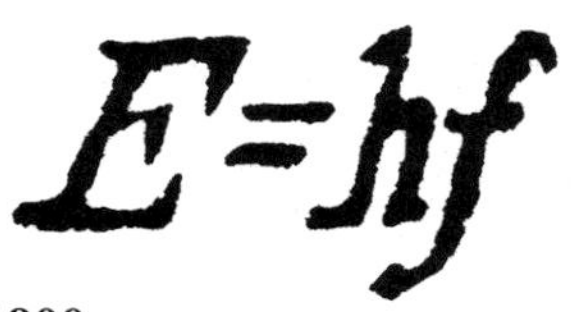

1900
Formula for Max Planck's law of electromagnetic radiation, or quantum theory.

1900 • British nuclear physicist Sir Ernest Rutherford (Lord Rutherford of Nelson) names beta rays (later known as beta particles). English physicist Sir Owen Willans Richardson discovers that heated metals tend to emit electrons (thermionic emission). Planck's radiation law stated by German physicist Max Ernst Ludwig Planck. Element radon discovered by German physicist Friedrich Ernst Dorn

1901 • Naming of element europium by Eugène-Anatole Demarçay. Dutch chemist Jacobus van't Hoff receives the first Nobel Prize in chemistry for work on laws of chemical dynamics and osmotic pressure. He pioneers the measurement and study of the rates and mechanisms of chemical reactions under different physical conditions. He applies thermodynamics to chemical reactions. Adrenaline isolated by Japanese chemist Jokichi Takamine

1902 • Czech scientist Bohuslav Brauner suggests that rare earth elements should be placed together in one space of the periodic table between lanthanum and tantalum. German organic chemist Emil Fischer receives the Nobel Prize in chemistry for work on sugar and purine syntheses

1903 • Hardening of fats by hydrogenation discovered. Industrial production of viscose begins. Swedish physical chemist Svante Arrhenius receives the Nobel Prize in chemistry for work on the theory of electrolytic dissociation

1904 • Scottish chemist William Ramsay receives the Nobel Prize in chemistry for work on the discovery of inert gases in the air and their locations in the periodic table

1905
Portrait of Albert Einstein, who devised the theory of relativity.

1905 • Special theory of relativity devised by German-born physicist Albert Einstein, including formula $E = mc^2$. German organic chemist Adolf von Baeyer receives the Nobel Prize in chemistry for work in organic dyes (particularly for his synthesis of indigo) and hydroaromatic compounds

1906 • French chemist Henri Moissan receives the Nobel Prize in chemistry for work on the isolation of fluorine and the adaptation of the electric furnace. Formulation of the third law of thermodynamics (entropy as a measure of disorder is a function of temperature) by German physical chemist Walther Hermann Nernst

1907 • Element lutetium discovered by French chemist Georges Urbain. Element ytterbium discovered by Swiss chemist Jean-Charles de Marignac. German chemist Eduard Buchner receives the Nobel Prize in chemistry for the discovery of cell-free fermentation

1908 • Paschen series of lines discovered by physicist Louis Paschen. British nuclear physicist Sir Ernest Rutherford receives the Nobel Prize in chemistry for work in atomic disintegration and the chemistry of radioactive substances

1909 • Haber-Bosch ammonia synthesis process in operation. Danish biochemist Søren Peter Lauritz Sørensen describes the effect of hydrogen ion concentration on enzyme activity and proposes the use of a negative logarithm of this concentration as a measure of acidity and alkalinity. A book about hydrogen ion concentration by German medical chemist Leonor Michalis makes the pH scale more widely known. German physical chemist Friedrich Wilhelm Ostwald receives the Nobel Prize in chemistry for work in catalysis, the principles of equilibria, and rates of reaction. DNA and RNA discovered by American (formerly Russian) biochemist Phoebus Levene. The first synthetic rubber is invented by German scientist Karl Hoffman

1910 • Existence of isotopes confirmed by English physicist Sir Joseph John Thomson. German organic chemist Otto Wallach receives the Nobel Prize in chemistry for work in alicyclic compounds. The first specific antibacterial agent (salvarsan, for treatment of syphilis) discovered by German founder of chemotherapy, Paul Ehrlich. Production of

Bakelite, a phenol-formaldehyde resin. This had been discovered by Belgian-American chemist Leo H. Baekeland (1863–1944) and was widely used in industry

1911 • French physicist Marie Curie receives the Nobel Prize in chemistry for work on the discovery of radium and polonium and the isolation and study of radium. Nuclear atom concept proposed by British nuclear physicist Sir Ernest Rutherford (Lord Rutherford of Nelson)

1912 • French organic chemist François Auguste Victor Grignard receives the Nobel Prize in chemistry for the discovery of Grignard reagents, which are useful in organic synthesis. French chemist Paul Sabatier receives the Nobel Prize in chemistry (with Grignard) for work on finding methods of catalytic hydrogenation of organic compounds. X-ray crystallography technique discovered by German theoretical physicist Max von Laue

1913 • British scientist Henry Moseley confirms the existence of exactly 14 rare-earth elements. Discovery of Stark effect, concerning the splitting of spectrum lines, by German physicist Johannes Stark. English chemist Frederick Soddy coins the term *isotope*. Quantum theory and Bohr atomic model proposed by Danish physicist Niels Bohr. Swiss chemist Alfred Werner receives the Nobel Prize in chemistry for work in bonding of atoms within molecules. Ozone layer in the Earth's atmosphere discovered by French scientist Charles Fabry. Vitamin A isolated by American scientist Elmer McCollum. Trial industrial plant for ammonia synthesis is built at Ludwigshafen

1914 • Name *proton* given to the positively charged nucleus of the hydrogen atom by British nuclear physicist Sir Ernest Rutherford (Lord Rutherford of Nelson). American chemist Theodore Richards receives the Nobel Prize in chemistry for work on the accurate determination of the atomic masses of many elements

1915 • General theory of relativity conceived by German-born physicist Albert Einstein. German organic chemist Richard Willstätter receives the Nobel Prize in chemistry for research into plant pigments, especially chlorophyll

1916 • American chemist Gilbert N. Lewis (1875–1946) introduces the concept of the shared pair of electrons in a chemical bond

1917 • Introduction of microanalytical methods by Austrian organic chemist Fritz Pregl. Discovery of element protoactinium by German chemist Otto Hahn and Austrian physicist Lise Meitner

1918 • German chemist Fritz Haber receives the Nobel Prize in chemistry for work on the synthesis of ammonia from its constituent elements

1919 • First artificial atomic fission achieved by British nuclear physicist Sir Ernest Rutherford (Lord Rutherford of Nelson)

1920 • German physical chemist Walther Hermann Nernst receives the Nobel Prize in chemistry for work in thermochemistry (the third law of thermodynamics)

1921 • English chemist Frederick Soddy receives the Nobel Prize in chemistry for work in radioactive substances, especially isotopes. Danish physicist Niels Bohr produces a detailed picture of the distribution of electronic shells

1922 • Discovery of the use of brown platinum oxide as a catalyst in hydrogenation by American chemist Roger Adams. English chemist Francis Aston receives the Nobel Prize in chemistry for work in mass spectrometry of isotopes of radioactive elements and the enunciation of the whole-number rule. Aston invents the mass spectrograph

1923 • American physicist Arthur Holly Compton coins word *photon*. Element hafnium discovered by Dutch physicist Dirk Coster and Hungarian chemist George de Hevesy.

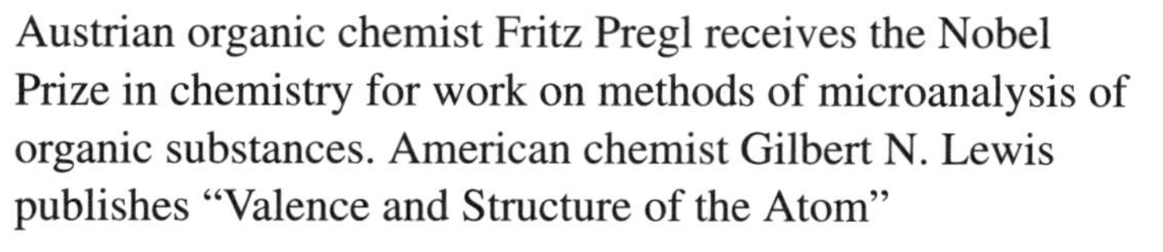

Austrian organic chemist Fritz Pregl receives the Nobel Prize in chemistry for work on methods of microanalysis of organic substances. American chemist Gilbert N. Lewis publishes "Valence and Structure of the Atom"

1924 • Discovery of wave nature of electrons by French physicist Prince Louis Victor de Broglie, later Duke de Broglie

1925
Wolfgang Pauli's "exclusion principle" explains the electronic structure of atoms.

1925 • Element rhenium discovered by German chemists Walter Noddack, Ida Tacke, and Otto C. Berg. Auger effect discovered by French physicist Pierre Auger. Austrian chemist Richard Zsigmondy receives the Nobel Prize in chemistry for work on elucidating the heterogeneity of colloids. Pauli "exclusion principle" (no two electrons can be in the same energy state) propounded by Austrian-born Swiss physicist Wolfgang Pauli. German physicist Friedrich Hund establishes "Hund's Rule"

1926 • Invention of Bernal chart by Irish physicist John Desmond Bernal to assist with the analysis of crystal structures. Swedish chemist Theodor Svedberg receives the Nobel Prize in chemistry for his investigation of dispersed systems. Isolation of urease by American biochemist James Sumner. German philosopher and physicist Werner Heisenberg develops "new quantum mechanics." Hypothesis of wave mechanics put forward by Austrian physicist Erwin Schrödinger (1887–1961) to study electrons in an atom

1927
Heinrich Wieland researches into bile acids.

1927 • Electron diffraction of crystals discovered independently by American physicist Clinton Joseph Davisson and English physicist George Paget Thomson. German organic chemist Heinrich Wieland receives the Nobel Prize in chemistry for his research on the constitution of bile acids and related substances

1927–28 • Hund-Mulliken interpretation of molecular spectra. This gives a description of the molecular orbital theory of bonding

1928 • By 1928 quantum physicists show that the orbital closest to the nucleus (s orbital) is spherical. The next three electron energy shells (p orbitals) are seen as having the shape of dumbbells along the three coordinate axes. Polymethyl methacrylate (Perspex) invented. German organic chemists Otto Diels (1876–1954) and Kurt Alder (1902–58) develop an addition reaction in which double-bonded dienes (compounds containing two double bonds separated by a single bond) are transformed into cyclic compounds. This is known as the Diels-Alder reaction and it is important for the synthesis of a range of natural products. Concept of parity of atomic states developed by Hungarian-American physicist Eugene Paul Wigner. German organic chemist Adolf Windaus receives the Nobel Prize in chemistry for his research on the constitution of sterols and related vitamins. Penicillin discovered by the chance exposure of a culture of staphylococci by Scottish biochemist Sir Alexander Fleming. Vitamin C discovered by American scientists Charles Glen King and Albert von Szent-Györgi (Hungarian-born)

1929 • German-born physicist Albert Einstein develops a unified field theory. English biochemist Sir Arthur Harden and Swedish chemist Hans von Euler-Chelpin receive the Nobel Prize in chemistry for work on the fermentation of sugar, and fermentative enzymes

1930 • Development of the Houdry petroleum cracking process by French chemical engineer Eugène Houdry. (Houdry moved to the United States to obtain finance and later became a U.S. citizen.) German organic chemist Hans Fischer receives the Nobel Prize in chemistry for his analysis of hem (the iron-bearing group in hemoglobin) and chlorophyll and the synthesis of hemin (a compound of hem)

1931 • Formulation of nylon, the first all-synthesized fiber (marketed by Du Pont in 1938), by Wallace H. Carothers. Carothers also studies the mechanism of polymerization and shows that it principally takes place by either addition or condensation reactions. Electron microscope invented by German scientists Ernst Ruska and Max Knoll. German chemical engineer Carl Bosch and German industrial chemist Friedrich Bergius receive the Nobel Prize in chemistry for their work on the invention and development of chemical high-pressure methods. Bergius uses these methods in the conversion of wood to coal. Bosch applies the methods to the synthesis of ammonia. Vitamin A structure identified by Russian-born Swiss organic chemist Paul Karrer. Erich Hückel (1896–1980) suggests that the stability of the benzene ring is caused by six sigma bonds in the plane of the ring and six pi electrons in orbits above and below the plane

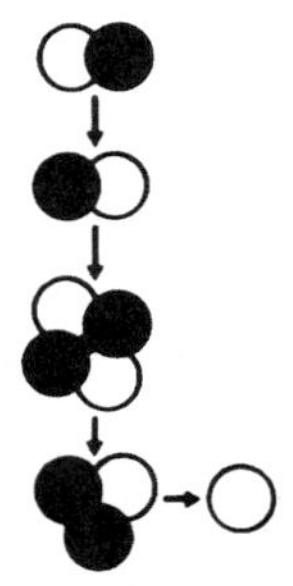

1932
James Chadwick's neutron.

1932 • First nuclear reaction to result from the bombardment of an element by artificially accelerated particles achieved by English physicist Sir John Cockcroft and Irish physicist Ernest Walton. Neutron discovered by English physicist Sir James Chadwick. American physical chemist Irving Langmuir receives the Nobel Prize in chemistry for his discoveries and investigations in surface chemistry. Positron discovered by American physicist Carl David Anderson. Urea cycle discovered by German biochemist Hans Krebs. Bactericidal properties of the red dye "Prontosil" (sulfamido chrysoidine) are recognized

1934
Enrico Fermi splits the nucleus of an atom of uranium–235.

1933 • Polythene developed. Magnetic characteristics and wave aspects of molecular beams demonstrated by German-American physicist Otto Stern

1934 • Mass of a neutron determined by English physicist Sir James Chadwick and Austrian-American physicist Maurice Goldhaber. Artificial radioactivity achieved for first time by French physicists Irene and Frédéric Joliot-Curie. Nucleus of atom split by Italian-American physicist Enrico Fermi. American chemist Harold Urey receives the Nobel Prize in chemistry for his discovery of deuterium (heavy hydrogen)

1935 • French physicists Irène and Frédéric Joliot-Curie receive the Nobel Prize in chemistry for their work on the synthesis of new radioactive elements

1936 • Dutch physicist and chemist Peter Debye receives the Nobel Prize in chemistry for his discoveries about molecular structures by studying dipole moments and the diffraction of X rays and electrons in gases. Development of electrophoresis by Swedish biochemist Arne Tiselius

1937 • "Muon" particle discovered by American physicist Carl David Anderson. Element technetium, the first element produced artificially, discovered by French physicist Carlo Perrier and Italian-born American Emilio Segrè. English chemist Sir Walter Norman Haworth receives the Nobel Prize in chemistry for his work in the study of carbohydrates and ascorbic acid (vitamin C). Swiss organic chemist Paul Karrer receives the Nobel Prize in chemistry for work in the study of carotenoids, flavins, retinol (vitamin A) and riboflavin (vitamin B_2)

1938 • PTFE (polytetraflouroethylene) invented in the United States. Discovery of chain reaction nuclear fission by German chemist Otto Hahn and Austrian physicist Lise Meitner. Technique of magnetic resonance developed by American physicist Isidor Isaac Rabi. German chemist Richard Kuhn is awarded the Nobel Prize in chemistry for work on carotenoids and research into vitamins but returns it. Nuclear fission discovered by German scientists Otto Hahn and Fritz Strassman

1939 • American physical organic chemist Linus Pauling publishes *Nature of the Chemical Bond*. Magnetic movement of a neutron calculated by American physicist Felix Bloch. Discovery of element francium by French physicist Marguerite Perey. German scientist Adolf Butenandt is awarded the Nobel Prize in chemistry for work in the field of sex hormones but declines to accept. Croatian-

1939
Linus Pauling publishes Nature of the Chemical Bond.

born Swiss chemist Leopold Ruzicka receives the Nobel Prize in chemistry for work on polymethylenes and higher terpenes

1940 • Discovery of element astatine by American physicists Emilio Segrè, Dale Corson, and K. R. Mackenzie. Discovery of element plutonium by American physicist Glenn T. Seaborg and colleagues. Discovery of element neptunium by American physicists Edwin H. McMillan and Phillip H. Abelson. DDT invented by Swiss scientist Paul Müller

1941 • Terylene synthetic fiber invented by English chemists John Whinfield and J. T. Dickson

1942 • First controlled chain reaction in a uranium and graphite pile created by Italian-American physicist Enrico Fermi

1943 • Hungarian scientist George de Hevesy receives the Nobel Prize in chemistry for work on the use of isotopes as tracers in chemical processes

1944 • Discovery of elements americium and curium by American physicist Glenn T. Seaborg and colleagues. German chemist Otto Hahn receives the Nobel Prize in chemistry for the discovery of nuclear fission. Paper chromatography invented by English biochemists Archer John Martin and Richard Synge, a technique for separating and identifing individual amino acids in a mixture

1946
Wendell Stanley works on enzymes and virus proteins.

1945 • Finnish scientist Artturi Virtanen receives the Nobel Prize in chemistry for work in agriculture and nutrition, especially the preservation of fodder

1946 • American biochemist James Sumner receives the Nobel Prize in chemistry for his discovery of the crystallization of enzymes. American biochemists John Northrop and Wendell Stanley receive the Nobel Prize in chemistry for their work on the preparation of pure enzymes and virus proteins

1947 • First true meson (pi-meson or pion) discovered by English physicist Cecil Frank Powell by investigating cosmic radiation at high altitudes. Element promethium discovered by American chemists J. A. Marinsky, L. E. Glendenin, and C. D. Coryell. English organic chemist Sir Robert Robinson receives the Nobel Prize in chemistry for work on the investigation of biologically important plant products, especially alkaloids

1948 • Shell model of atomic nucleus advanced by American physicist Maria Goeppert Mayer and German physicist Johannes Hans Daniel, who independently introduce the concept of magic numbers. (These are the numbers 2, 8, 20, 28, 50, 82, or 126. If a nucleus has a magic number of either protons or neutrons, it is more than usually stable.) Swedish biochemist Arne Tiselius receives the Nobel Prize in chemistry for work on research into electrophoresis and adsorption analysis, and discoveries concerning serum proteins. Quantum electrodynamics invented by American theoretical physicist Richard Feynman, American physicist Seymour Schwinger, and Japanese physicist Sin-itiro Tomonaga

1949 • Discovery of element berkelium by American physicist Glenn T. Seaborg and colleagues. American physical chemist William Giauque receives the Nobel Prize in chemistry for his work in the field of chemical thermodynamics, particularly on the behavior of substances at very low temperatures

1950s • Two main classes of elementary particles are identified: hadrons (including nucleons, mesons, and hyperons) and leptons (including electrons, neutrinos, and muons)

1950 • Discovery of element californium by American physicist Glenn T. Seaborg and colleagues. Acrylic fiber invented. German organic chemists Otto Diels and Kurt Alder receive the Nobel Prize in chemistry for work on the discovery and development of diene synthesis

1951 • American physicists Edwin McMillan and Glenn T. Seaborg receive the Nobel Prize in chemistry for their discovery of, and work in, the chemistry of the transuranic elements

1952 • Discovery of K meson (or kaon) and lambda particle by Polish physicists Marian Danysz and Jerzy Pniewski. Discovery of element einsteinium by American physicist Albert Ghiorso and colleagues. English biochemists Archer John Martin and Richard Synge receive the Nobel Prize in chemistry for work on the development of partition chromatography. Element fermium is discovered in the remains of the first thermonuclear explosion by American physicist Albert Ghiorso and colleagues

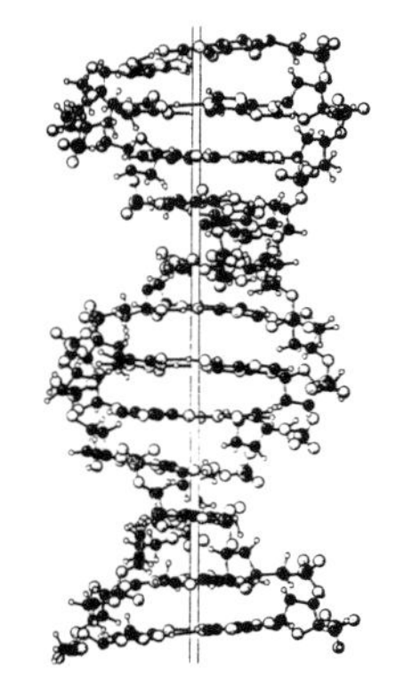

1953
Francis Crick and James Watson receive Nobel Prize for discovering structure of DNA.

1953 • Bubble chamber, for detecting ionizing radiation, invented by American physicist Donald Arthur Glaser. Maser, forerunner of laser, produced by American physicist Charles Townes. Hermann Staudinger, German founder of polymer chemistry, receives the Nobel Prize in chemistry for his discoveries in macromolecular chemistry. The double-helix structure of DNA is discovered by molecular biologists Francis Crick (English) and James Watson (American). They, together with Maurice Wilkins (English), receive the Nobel Prize in physiology or medicine in 1962 for this discovery and its significance in the replication and transfer of genetic information

1954 • First particle accelerator built. American physical organic chemist Linus Pauling receives the Nobel Prize in chemistry for work on the study of the nature of chemical bonds, especially in complex substances

1955 • Invention of field ion microscope, the first device to yield images of individual atoms, by American physicist Erwin Wilhelm Mueller. Two types of K mesons are detected with differing modes of decay. Discovery of element mendelevium by American physicist Albert Ghiorso and colleagues. Antiprotons discovered; they are negatively charged particles that have the mass of protons. American biochemist Vincent du Vigneaud receives the Nobel Prize in chemistry for investigations into biochemically important sulfur compounds and the first synthesis of a polypeptide hormone. Vitamin B composition identified by English chemist Dorothy Crowfoot Hodgkin

1956 • American physicists Frederick Reines and Clyde Lorrain Cowan discover antineutrinos. English physical chemist Sir Cyril Hinshelwood and Russian scientist Nikolai Semenov receive the Nobel Prize in chemistry for their investigation of the mechanism of chemical reactions

1957 • Scottish bio-organic chemist Alexander Todd (Baron Todd of Trumpington) receives the Nobel Prize in chemistry for work in nucleotides and nucleotide coenzymes. American biochemist John Sheenan synthesizes penicillin

1958 • Discovery of element nobelium by American physicist Albert Ghiorso and colleagues. English molecular biochemist Frederick Sanger receives the Nobel Prize in chemistry for the determination of the structure of proteins, particularly insulin

1959 • Japanese physicists Saburo Fukui and Shotaro Miyamoto invent spark chamber to detect ionizing particles selectively. Czech physical chemist Jaroslav Heyrovsky receives the Nobel Prize in chemistry for work on the discovery and development of polarographic methods of chemical analysis

1960s • Advent of high-resolution mass spectrometers and nuclear magnetic resonance (NMR) spectroscopy

1960 • American chemist Willard Libby receives the Nobel Prize in chemistry for the development of radiocarbon dating in archaeology, geology, and geography. Messenger RNA discovered by South African molecular biologist Sydney Brenner and French molecular biologist François Jacob

1961 • Record set by Russian military scientists for largest nuclear explosion when they test a 58-megaton weapon. Elementary particles called hadrons classified by American physicist Murray Gell-Mann in a system he calls Eightfold Way. American physicist Albert Ghiorso and colleagues discover element lawrencium. American chemist Melvin Calvin receives the Nobel Prize in chemistry for his study of the assimilation of carbon dioxide by plants

1962 • British physicist Heinz London develops a technique for inducing very low temperatures with mixture of helium-3 and helium-4. Canadian chemist Neil Bartlett combines noble gas xenon with platinum fluoride to produce xenon fluoroplatinate, the first-known case of a noble gas bonding with another element to form a compound. Austrian-born British molecular biologist Max Perutz and English molecular biologist Sir John Kendrew receive the Nobel Prize in chemistry for their work on the determination of the structures of globular proteins

1963
Murray Gell-Mann suggests the existence of the quark.

1963 • German chemist Karl Ziegler and Italian chemist Giulio Natta receive the Nobel Prize in chemistry for work on the chemistry and technology of high-melting-point polymers. The subatomic particle, the quark, first suggested by American physicists Murray Gell-Mann and George Zweigus. Carbon fiber invented by English scientist Leslie Phillips

1964 • English chemist Dorothy Crowfoot Hodgkin receives the Nobel Prize in chemistry for her work on the crystallographic determination of the structures of biochemical compounds, particularly penicillin and cyanocobalamin (vitamin B_{12})

1965 • American organic chemist Robert Woodward receives the Nobel Prize in chemistry for work in the field of organic synthesis

1966 • American chemist Robert Mulliken receives the Nobel Prize in chemistry for work on the molecular orbital theory of chemical bonds and structures

1967 • Discovery of element dubnium by Russian scientists. German chemist Manfred Eigen, British chemist Ronald Norrish, and British chemist George Porter (Baron Potter of Luddenham) receive the Nobel Prize in chemistry for their work on the investigation of rapid chemical reactions by means of very short pulses of energy. Ronald Norrish and George Porter develop the technique of flash photolysis. American physical organic chemist Linus Pauling, in *The Chemical Bond: A Brief Introduction to Modern Structural Chemistry,* includes the principles of the molecular orbital theory while stating that, for introductory teaching and the consideration of the ground states of molecules, the valence bond theory is still preferable

1968 • American chemist Lars Onsager receives the Nobel Prize in chemistry for the discovery of reciprocal relations, which are fundamental for the thermodynamics of irreversible processes

1969 • Discovery of element rutherfordium by American physicist Albert Ghiorso and colleagues. British organic chemist Sir Derek Barton and Norwegian chemist Odd Hassel receive the Nobel Prize in chemistry for work on the concept of conformation and its applications. Structure of insulin discovered by English chemist Dorothy Crowfoot Hodgkin

1969
Dorothy Crowfoot Hodgkin discovers structure of insulin.

1970 • Argentinian chemist Luis Federico Leloir receives the Nobel Prize in chemistry for the discovery of sugar nucleotides and determining their role in carbohydrate biosynthesis

1971 • Protein is obtained from hydrocarbons. Canadian chemist Gerhard Herzberg receives the Nobel Prize in chemistry for research into the electronic structure and geometry of molecules, particularly free radicals

1972 • American chemists Christian Anfinsen, Stanford Moore, and William Stein receive the Nobel Prize in chemistry for work in determining the structure of amino acids and research into the biological activity of the enzyme ribonuclease

1973 • West German chemist Ernst Fischer and British chemist Geoffrey Wilkinson receive the Nobel Prize in chemistry for their work on the chemistry of organometallic sandwich compounds

1974 • Element seaborgium discovered by American physicists Albert Ghiorso and colleagues. American chemist Paul Flory receives the Nobel Prize in chemistry for his studies of the physical chemistry of macromolecules

1975 • Australian chemist Sir John Cornforth receives the Nobel Prize in chemistry for work in the stereochemistry of enzyme-catalyzed reactions. Swiss chemist Vladimir Prelog receives the Nobel Prize in chemistry for work in the stereochemistry of organic molecules and their reactions

1976 • Two unmanned NASA probes, *Viking 1* and *Viking 2*, touch down on the surface of the planet Mars and conduct the first chemical analysis of the surface of that planet. The prime mission objective for the landing craft is to determine if life is or ever has been present on the planet, using chemical analysis of the soil and atmosphere. Probe telemetry reports high concentrations of iron in the soil but no sign of life. American chemist William Lipscomb receives the Nobel Prize in chemistry for the study of the structure and chemical bonding of boranes (compounds of boron and hydrogen). Recombinant DNA technique identified by American scientists Stanley Cohen and Herbert Boyer

1977 • Belgian chemist Ilya Prigogine receives the Nobel Prize in chemistry for research into the thermodynamics of irreversible and dissipative processes

1978 • British chemist Peter Mitchell receives the Nobel Prize in chemistry for the formulation of a theory of biological energy transfer and chemiosmotic theory

1979 • American chemist Herbert Brown and German chemist Georg Wittig receive the Nobel Prize in chemistry for their work on uses of boron and phosphorus compounds in organic syntheses

1980 • Development of scanning tunneling microscope, which can produce images of individual atoms on the surface of a material. American chemist Paul Berg receives the Nobel Prize in chemistry for his work on the biochemistry of nucleic acids, especially recombinant DNA. American chemist Walter Gilbert and English molecular biochemist Frederick Sanger receive the Nobel Prize in chemistry for their work on the base sequences in nucleic acids

1981 • Physicists in Darmstadt, Germany, confirm the existence of element bohrium, after Russian scientists originally report its discovery in 1976. Japanese chemist Kenichi Fukui and American chemist Roald Hoffmann receive the Nobel Prize in chemistry for their work on theories concerning chemical reactions

1982 • Unmanned Russian spacecraft *Venera 13* dispatches a landing craft to the surface of the planet Venus, which conducts the first successful chemical analysis of the surface of this hostile environment. Analysis of a surface sample by the lander's X-ray fluorescence spectrometer classifies the material as melanocratic alkaline gabbroids. Discovery of element meitnerium by German scientists. British chemist Aaron Klug receives the Nobel Prize in chemistry for work on the determination of crystallographic electron microscopy and the structure of biologically important nucleic acid protein complexes

1983 • American chemist Henry Taube receives the Nobel Prize in chemistry for the study of electron-transfer reactions in inorganic chemical reactions

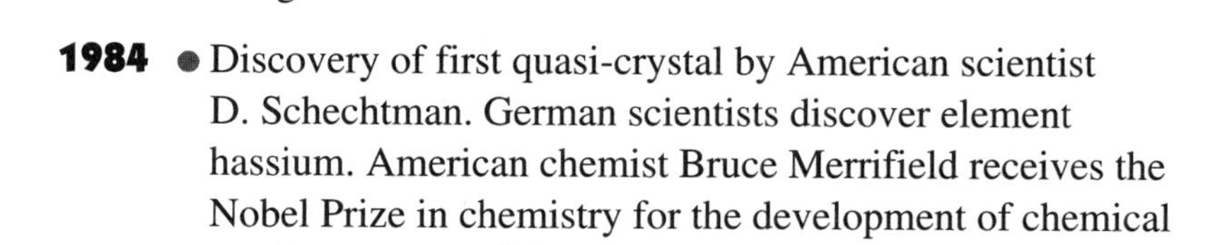

1984 • Discovery of first quasi-crystal by American scientist D. Schechtman. German scientists discover element hassium. American chemist Bruce Merrifield receives the Nobel Prize in chemistry for the development of chemical syntheses on a solid matrix

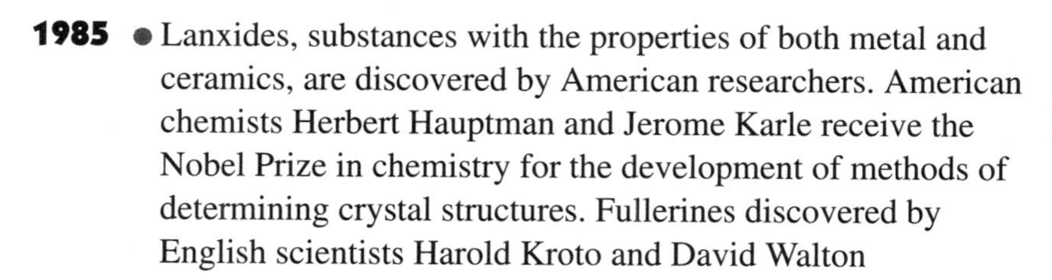

1985 • Lanxides, substances with the properties of both metal and ceramics, are discovered by American researchers. American chemists Herbert Hauptman and Jerome Karle receive the Nobel Prize in chemistry for the development of methods of determining crystal structures. Fullerines discovered by English scientists Harold Kroto and David Walton

1987
Donald Cram, who worked on understanding how specific molecules recognize each other.

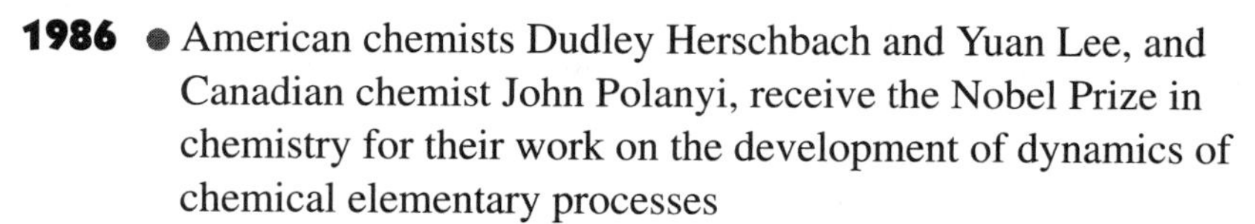

1986 • American chemists Dudley Herschbach and Yuan Lee, and Canadian chemist John Polanyi, receive the Nobel Prize in chemistry for their work on the development of dynamics of chemical elementary processes

1987 • American chemists Donald Cram and Charles Pedersen and French chemist Jean-Marie Lehn receive the Nobel Prize in chemistry for work on the development of molecules with highly selective structure-specific interactions

1988 • German chemists Johann Deisenhofer, Robert Huber, and Hartmut Michel receive the Nobel Prize in chemistry for their discovery of the three-dimensional structure of the reaction center of photosynthesis

1989 • American chemists Sidney Altman and Thomas Cech receive the Nobel Prize in chemistry for their discovery of the catalytic function of RNA

1990 • American chemist Elias James Corey receives the Nobel Prize in chemistry for discovering new methods of synthesizing chemical compounds

1991 • American chemist Joel Hawkins corroborates existence of the buckyball molecule (or buckminsterfullerene), a form of pure carbon. Swiss chemist Richard Ernst receives the Nobel Prize in chemistry for improvements to the technology of nuclear magnetic resonance (NMR) imaging

1992 • American chemist Rudolph Marcus receives the Nobel Prize in chemistry for making theoretical discoveries concerning reduction and oxidation reactions

1993 • American chemist Kary Mullis receives the Nobel Prize in chemistry for the invention of the polymerase chain reaction technique for amplifying DNA. Canadian chemist Michael Smith receives the Nobel Prize in chemistry for his invention of techniques for splicing foreign genetic segments into an organism's DNA to modify the proteins produced

1994 • American chemist George Olah receives the Nobel Prize in chemistry for his development of a technique used to examine hydrocarbon molecules. The elements ununnilium and unununium are discovered. These are temporary names

1995 • American chemists F. Sherwood Rowland and Mario Molina, and Dutch chemist Paul Crutzen, receive the Nobel Prize in chemistry for their explanation of the chemical processes of the ozone layer

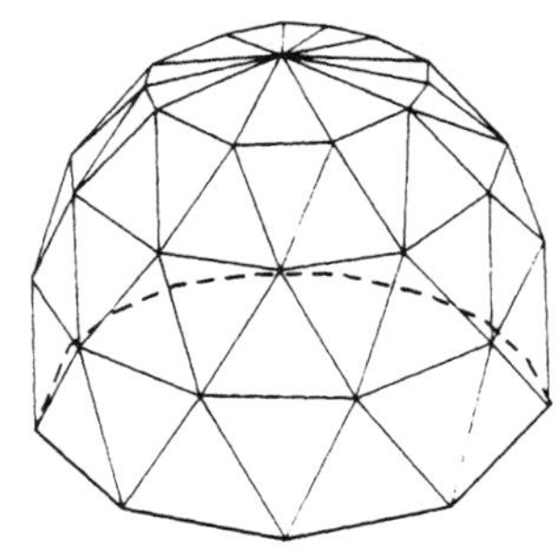

1996
Geodesic dome construction of Buckminster Fuller.

1996 • American chemists Robert Curl, Jr, and Richard Smalley, and British chemist Harold Kroto, receive the Nobel Prize in chemistry for their discovery of the buckminsterfullerene molecule, an entirely new kind of molecule to chemistry (named after Buckminster Fuller, designer of the geodesic dome, which is similar in appearance to the spherical structure of the molecule). The molecules are also known as either fullerenes or buckyballs. The element ununbium is discovered. This is a temporary name

1997 • British chemist John Walker, American chemist Paul Boyer, and Danish chemist Jens Skou receive the Nobel Prize in chemistry for their studies of the enzymes involved in the production of adenosine triphospate (ATP)

1998 • British chemist John A. Pople and American chemist Walter Kohn receive the Nobel Prize in chemistry for theoretical work on molecules. A spectrometer on board NASA's unmanned Moon probe *Lunar Prospector* detects possible traces of water at the Moon's north and south poles. This discovery is considered vital to the future of manned exploration of the Moon and the inner solar system

1999 • NASA deliberately crashes the unmanned *Lunar Prospector* probe into the surface of the Moon, hoping to create a dust plume that can be conclusively analyzed for the prescence of water on the lunar surface. Data from the world's most powerful telescopes and spectrometers trained on the impact site remain inconclusive

SECTION FOUR
CHARTS & TABLES

Elements

In the following table elements are listed by letter symbol. The list includes the atomic number, element name, and the atomic weight of each element.
* indicates the atomic weight of the isotope with the lowest known half-life.

Periodic table

1 H																	2 He
3 Li	4 Be											5 B	6 C	7 N	8 O	9 F	10 Ne
11 Na	12 Mg											13 Al	14 Si	15 P	16 S	17 Cl	18 Ar
19 K	20 Ca	21 Sc	22 Ti	23 V	24 Cr	25 Mn	26 Fe	27 Co	28 Ni	29 Cu	30 Zn	31 Ga	32 Ge	33 As	34 Se	35 Br	36 Kr
37 Rb	38 Sr	39 Y	40 Zr	41 Nb	42 Mo	43 Tc	44 Ru	45 Rh	46 Pd	47 Ag	48 Cd	49 In	50 Sn	51 Sb	52 Te	53 I	54 Xe
55 Cs	56 Ba	57-71 -	72 Hf	73 Ta	74 W	75 Re	76 Os	77 Ir	78 Pt	79 Au	80 Hg	81 Tl	82 Pb	83 Bi	84 Po	85 At	86 Rn
87 Fr	88 Ra	89-103 -	104 Rf	105 Db	106 Sg	107 Bh	108 Hs	109 Mt									
		57 La	58 Ce	59 Pr	60 Nd	61 Pm	62 Sm	63 Eu	64 Gd	65 Tb	66 Dy	67 Ho	68 Er	69 Tm	70 Yb	71 Lu	
		89 Ac	90 Th	91 Pa	92 U	93 Np	94 Pu	95 Am	96 Cm	97 Bk	98 Cf	99 Es	100 Fm	101 Md	102 No	103 Lr	

Letter symbol	Atomic number	Name	Atomic weight
Ac	89	actinium	227.0278*
Ag	47	silver	107.868
Al	13	aluminum	26.98154
Am	95	americium	243.0614*
Ar	18	argon	39.948
As	33	arsenic	74.9216
At	85	astatine	209.987*
Au	79	gold	196.9665
B	5	boron	10.81
Ba	56	barium	137.33
Be	4	beryllium	9.0128
Bh	107	bohrium	
Bk	97	berkelium	247.0703*
Bi	83	bismuth	208.9804
Br	35	bromine	79.904
C	6	carbon	12.011
Ca	20	calcium	40.08
Cd	48	cadmium	112.41
Ce	58	cerium	140.12
Cf	98	californium	251.0796*
Cl	17	chlorine	35.453
Cm	96	curium	247.0703*
Co	27	cobalt	58.9332
Cr	24	chromium	51.996
Cs	55	cesium	132.9054
Cu	29	copper	63.546
Db	105	dubnium	
Dy	66	dysprosium	162.5
Er	68	erbium	167.26
Es	99	einsteinium	254.088*
Eu	63	europium	151.96
F	9	fluorine	18.9984
Fe	26	iron	55.847
Fm	100	fermium	257.0951*
Fr	87	francium	223.0197*
Ga	31	gallium	69.72
Gd	64	gadolinium	157.25
Ge	32	germanium	72.59
H	1	hydrogen	1.0079
He	2	helium	4.0026
Hf	72	hafnium	178.49
Hg	80	mercury	200.59
Ho	67	holmium	164.9304
Hs	108	hassium	
I	53	iodine	126.9045
In	49	indium	114.82
Ir	77	iridium	192.22
K	19	potassium	39.0983
Kr	36	krypton	83.8
La	57	lanthanum	138.9055
Li	3	lithium	6.941
Lr	103	lawrencium	260.105*
Lu	71	lutetium	174.967
Md	101	mendelevium	258.099*
Mg	12	magnesium	24.305
Mn	25	manganese	54.938
Mo	42	molybdenum	95.94
Mt	109	meitnerium	
N	7	nitrogen	14.0067
Na	11	sodium	22.98977
Nb	41	niobium	92.9064
Nd	60	neodymium	144.24
Ne	10	neon	20.179
Ni	28	nickel	58.69
No	102	nobelium	259.101*
Np	93	neptunium	237.0482*
O	8	oxygen	15.9994
Os	76	osmium	190.2
P	15	phosphorus	30.97376
Pa	91	protoactinium	231.0359
Pb	82	lead	207.19
Pd	46	palladium	106.42
Pm	61	promethium	144.9128*
Po	84	polonium	208.9824*
Pr	59	praseodymium	140.9077
Pt	78	platinum	195.08
Pu	94	plutonium	244.0642*
Ra	88	radium	226.0254*
Rb	37	rubidium	85.4678
Re	75	rhenium	186.207
Rf	104	rutherfordium	
Rh	45	rhodium	102.9055
Rn	86	radon	222.0176*
Ru	44	ruthenium	101.07
S	16	sulfur	32.064
Sb	51	antimony	121.75
Sc	21	scandium	44.9559
Se	34	selenium	78.96
Sg	106	seaborgium	
Si	14	silicon	28.0855
Sm	62	samarium	150.36
Sn	50	tin	118.69
Sr	38	strontium	87.62
Ta	73	tantalum	180.9479
Tb	65	terbium	158.9254
Tc	43	technetium	96.9064*
Te	52	tellurium	127.6
Th	90	thorium	232.0381
Ti	22	titanium	47.88
Tl	81	thallium	204.383
Tm	69	thulium	168.9342
U	92	uranium	238.029*
Uub	112	ununbium	277
Uun	110	ununnilium	269
Uuu	111	unuuunium	272
V	23	vanadium	50.9415
W	74	tungsten	183.85
Xe	54	xenon	131.29
Y	39	yttrium	88.9059
Yb	70	ytterbium	173.04
Zn	30	zinc	65.381
Zr	40	zirconium	91.224

Elements by groups

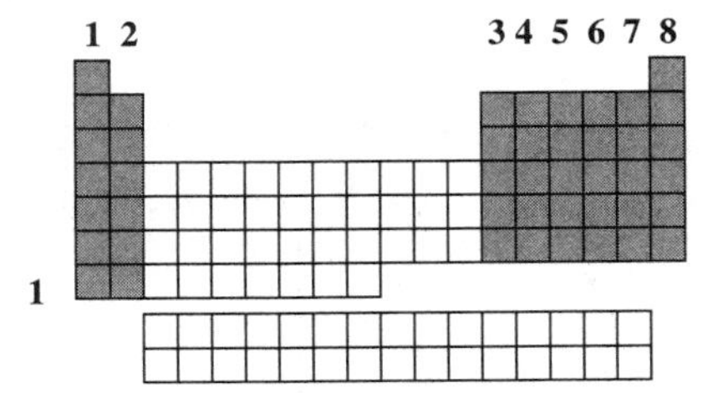

a
b
2
a group 4 element

1 The eight groups read downward.
2 These rank elements by number of electrons (**a**) in an atom's outer shell (**b**): Group 1, one electron; Group 2, two electrons, and so on through Group 8. Elements with the same number of outer shell electrons share similar properties. (Note: hydrogen fits no group, and helium, although in group 8, has only two electrons.)
Group 1 Alkali metals, the sodium family, with one electron in the outer shell. These are similar, very active metals.
Group 2 Alkaline-earth metals, the calcium family, with two electrons in the outer shell.
Group 3 Nonmetallic through metallic elements, with increasingly complex atoms. All have three electrons in the outer shell, and stable inner shells.
Group 4 Nonmetallic through metallic elements, also with increasingly complex atoms. All have four electrons in the outer shell, and stable inner shells.
Group 5 The nitrogen family – from nonmetallic nitrogen and phosphorus to metallic bismuth. All have five electrons in the outer shell, and stable inner shells.
Group 6 The oxygen family – from oxygen to metallic polonium. All have six electrons in the outer shell, and stable inner shells.
Group 7 The halogen family of active nonmetals. All have seven electrons in the outer shell, and stable inner shells.
Group 8 The inert gases. None chemically combines with any element. All (except helium) have eight electrons in the outer shell.

Electron arrangement of atoms

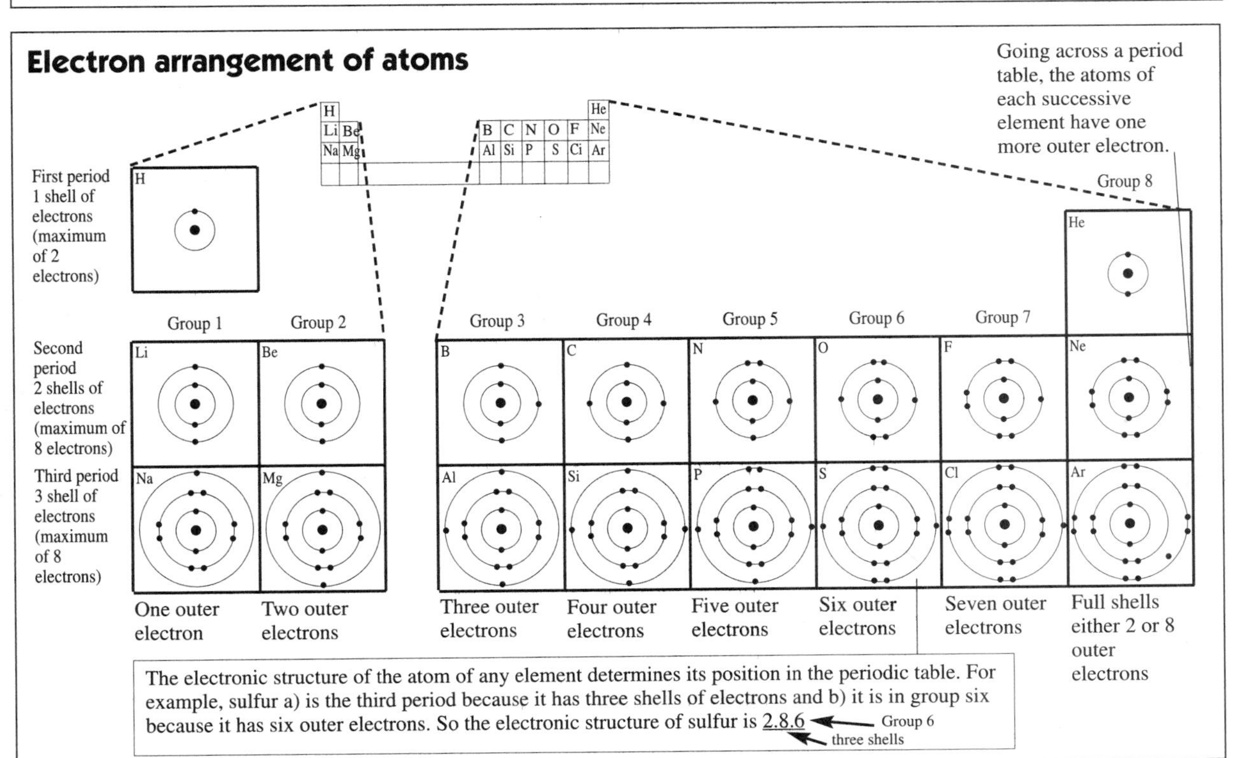

The electronic structure of the atom of any element determines its position in the periodic table. For example, sulfur a) is the third period because it has three shells of electrons and b) it is in group six because it has six outer electrons. So the electronic structure of sulfur is 2.8.6 ← Group 6 ← three shells

Metals and alloys

All but 25 of the known elements are metals. Metals are elements whose atoms can lose one or more electrons to form electrically positive ions. Most metals are good conductors of heat and electricity. They are malleable (can be beaten or rolled into a new shape) and ductile (can be pulled out into long wires). All metals are shiny, crystalline solids, except mercury, which is a liquid.

Activity series
Some metals form positive ions more easily than others, and so are more chemically active. Sixteen common metals are listed in the order of their activity. Lithium is the most active of all the metals, and gold is the least active.

Most active

lithium
potassium
calcium
sodium
magnesium
aluminum
zinc
chromium
iron
nickel
tin
lead
copper
silver
platinum
gold

Least active

Alloys
An alloy is a mixture of two or more metals. Here we list some everyday alloys, the metals from which they are made, and examples of their use.

Alloy	Metals	Examples of use
bronze	copper, tin	"copper" coins
brass	copper, zinc	doorhandles, buttons
cupronickel	copper, nickel	"silver" coins
pewter	tin, lead	tankards
stainless steel	iron, chromium, nickel	cutlery, pots, etc.
sterling silver	silver, copper	jewelry
9, 18, and 22 carat gold	gold, silver, copper	jewelry
dental amalgam	silver, tin, copper, zinc, mercury	filling cavities in teeth
solder	lead, tin	joining metals

Native metals
Only four of the least active metals – copper, silver, platinum, and gold – commonly occur in the Earth's crust as native metals (i.e. as free elements). All the others are found in compounds, called ores, which must be chemically treated to obtain the pure element.

Metalloids
These elements are "halfway" between metals and nonmetals. Depending on the way they are treated, they can act as insulators like nonmetals or conduct electricity like metals. This makes several metalloids extremely important as semiconductors in computers and other electronic devices. The eight metalloid elements are boron, silicon, germanium, arsenic, antimony, tellurium, polonium, and astatine.

Chemical reaction types

In a chemical reaction, molecules of a substance gain or lose atoms or atoms are rearranged. There are four main kinds of chemical reaction.
1 Combination: Two or more substances combine, forming a compound
2 Decomposition: A chemical compound breaks up into simpler substances
3 Replacement (substitution): A compound loses one or more atoms but gains other atoms instead
4 Double decomposition (double replacement): Two compounds decompose, exchanging atoms to form two new compounds

1

3

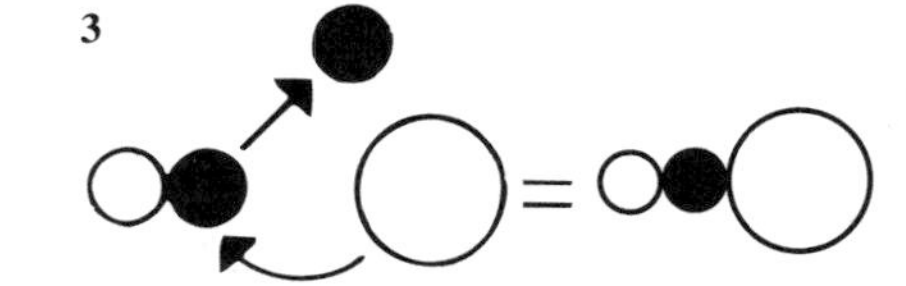

2

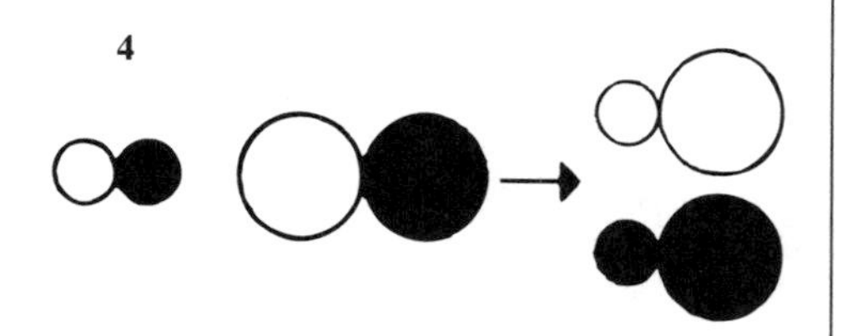

4

Chemicals from oil

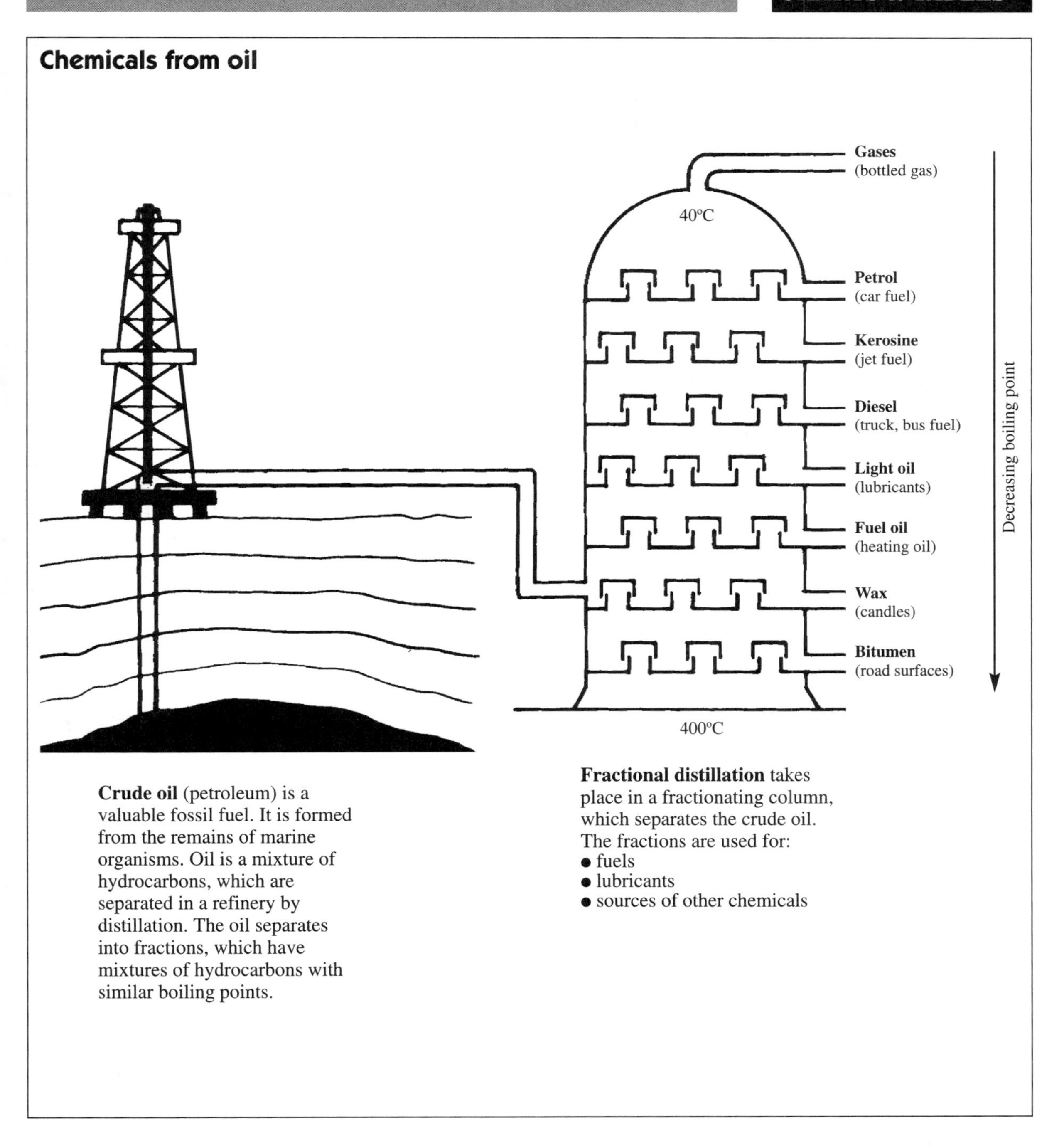

Crude oil (petroleum) is a valuable fossil fuel. It is formed from the remains of marine organisms. Oil is a mixture of hydrocarbons, which are separated in a refinery by distillation. The oil separates into fractions, which have mixtures of hydrocarbons with similar boiling points.

Fractional distillation takes place in a fractionating column, which separates the crude oil. The fractions are used for:

- fuels
- lubricants
- sources of other chemicals

Hydrocarbons

ALKENES These contain only carbon and hydrogen. There are four main types: Alkenes, Alkanes, Alkynes, and Aromatic hydrocarbons. Contain double bonds between carbon atoms.

Name	Molecular formula	Structural formula
ethene	C_2H_4 or $CH_2{=}CH_2$	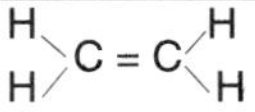 H₂C = CH₂ drawn as: H, H on C = C with H, H
propene	C_3H_6 or $CH_3CH{=}CH_2$	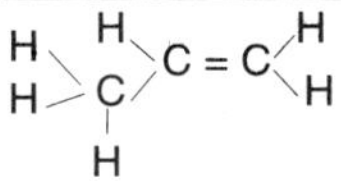 H, H, H on C, H (C–H below); C = C with H, H
butene	C_4H_8 or $CH_3CH_2CH{=}CH_2$ or $CH_3CH{=}CHCH_3$	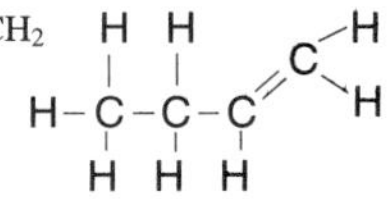 H H / H–C–C–C=C(H)(H) / H H H and H, CH_3 on C = C with H, CH_3 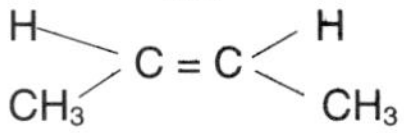

ALKANES Contain only single bonds.

Name	Molecular formula	Structural formula
methane	CH_4	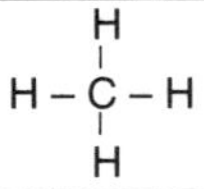 H H – C – H H
ethane	C_2H_6 or CH_3CH_3	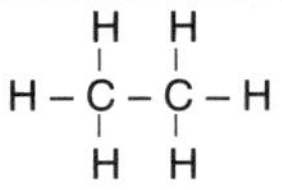 H H H – C – C – H H H
propane	C_3H_8 or $CH_3CH_2CH_3$	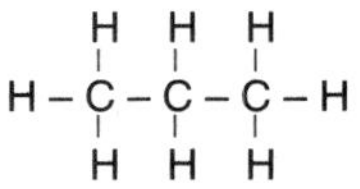 H H H H – C – C – C – H H H H
butane	C_4H_{10} or $CH_3CH_2CH_2CH_3$	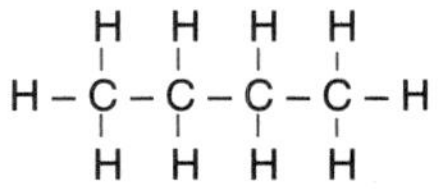 H H H H H – C – C – C – C – H H H H H
pentane	C_5H_{12} or $CH_3CH_2CH_2CH_2CH_3$	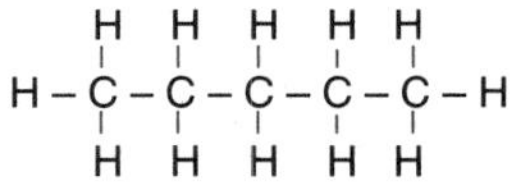 H H H H H H – C – C – C – C – C – H H H H H H

ALKYNES Contain triple bonds between carbon atoms.

Name	Molecular formula	Structural formula
ethyne	C_2H_2	 H – C ≡ C – H
propyne	C_3H_4	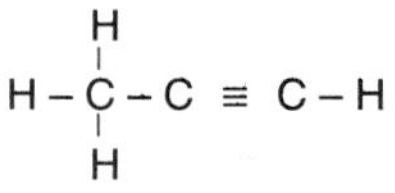 H H – C – C ≡ C – H H
butyne	C_4H_6	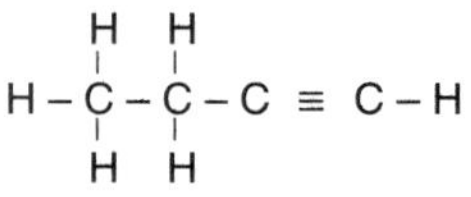 H H H – C – C – C ≡ C – H H H or H H H – C – C ≡ C – C – H H H 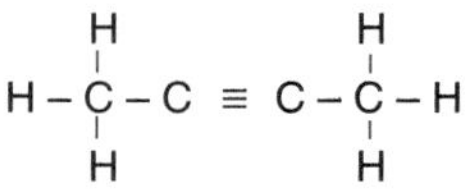

AROMATIC HYDROCARBONS Have six-sided rings with alternating double and single bonds.

Name	Molecular formula	Structural formula
benzene	C_6H_6	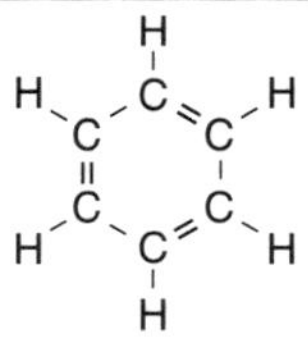 six-membered ring of C with alternating double bonds, each C bearing H
toluene	$C_6H_5CH_3$	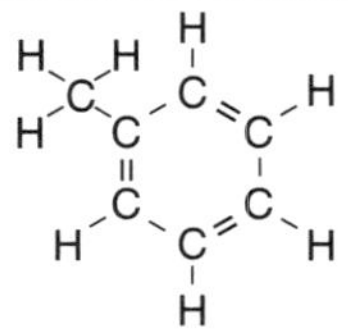 benzene ring with one H replaced by CH_3 (H, H, H on C)
naphthalene	$C_{10}H_8$	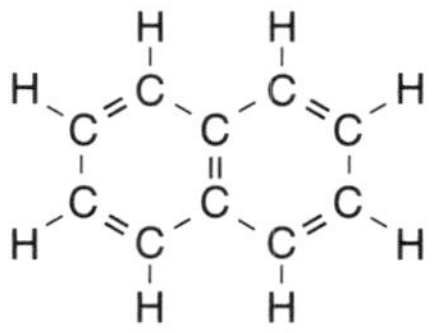 two fused six-membered C rings with alternating double bonds, eight H

Isomers

These are compounds with the same molecular formula but different structural formulas.

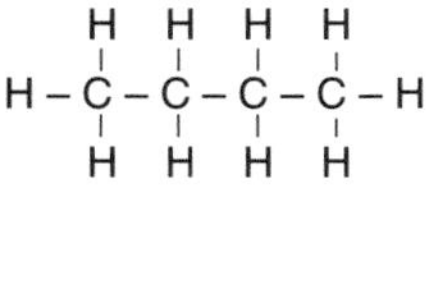

butane

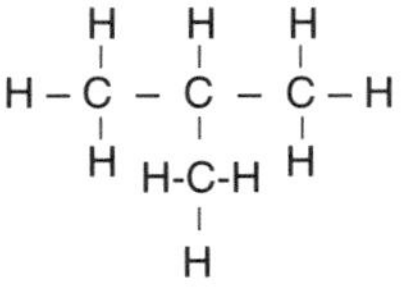

2 – methylopropane

structural isomers of the hydrocarbon C_4H_{10}

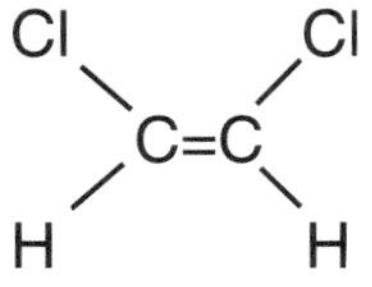

cis – 1, 2 dichloroethane

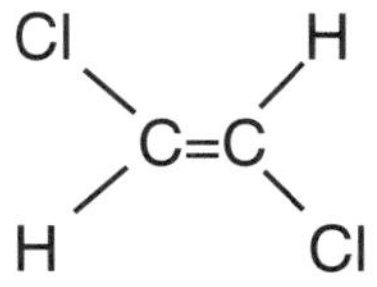

trans – 1,2 – dichloroethane

cis-trans isomers differing in arrangement about a double bond

---

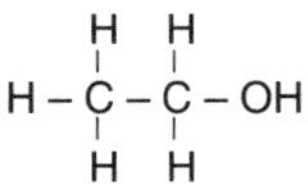

ethanol

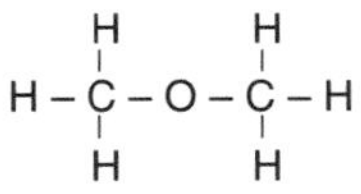

dimethyl ether

structural isomers differing in functional groups

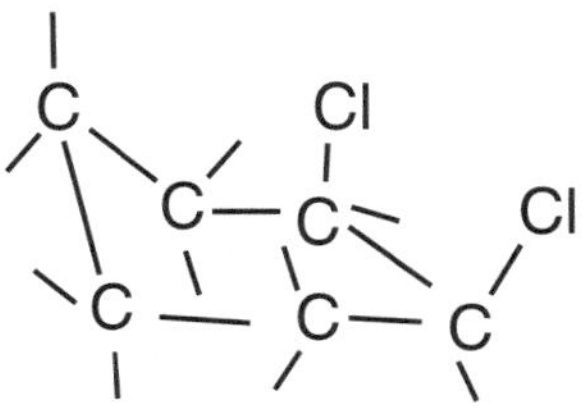

cis

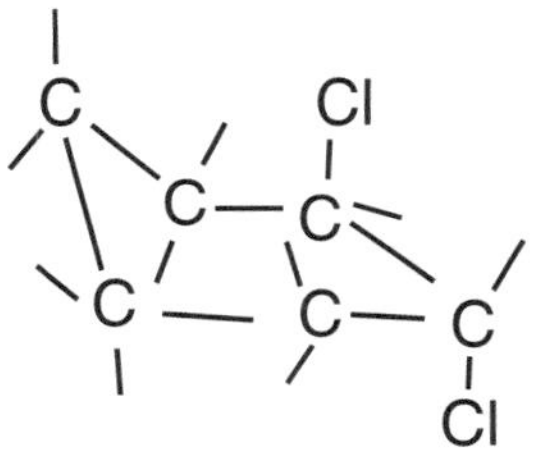

trans

cis-trans isomers differing in arrangement about a single bond with restricted rotation

---

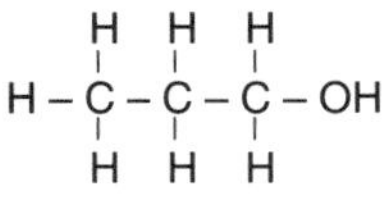

propan-1-ol

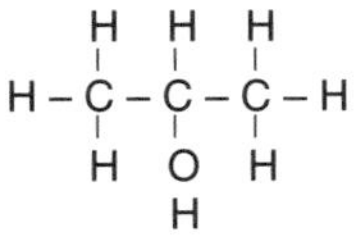

propan-2-ol

structural isomers differing in the position of the functional group

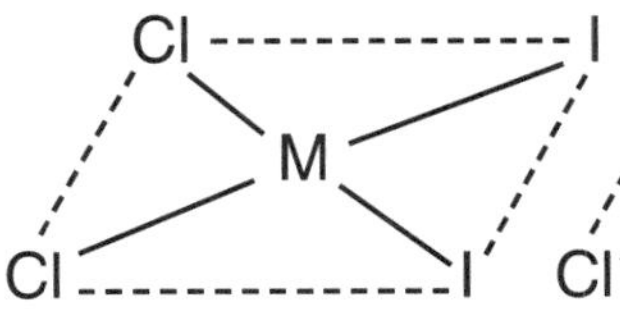

cis

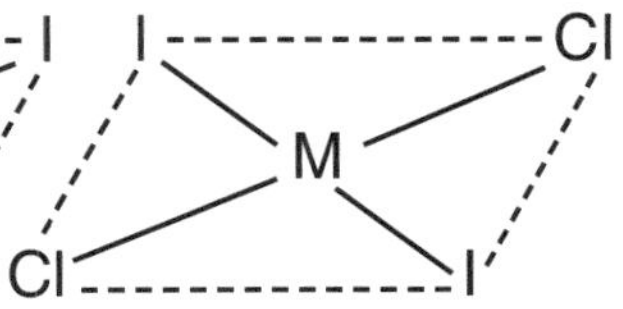

trans

cis-trans isomers differing in a planar metal complex

Polymers

Polymers
These are very large, usually long-chain, molecules made by linking together numbers of small molecules called monomers.

n A → – A – A – A – A

monomers polymer

Natural Polymers
These occur in all plants and animals.

Carbohydrates
In carbohydrates glucose is linked to make starch or cellulose.

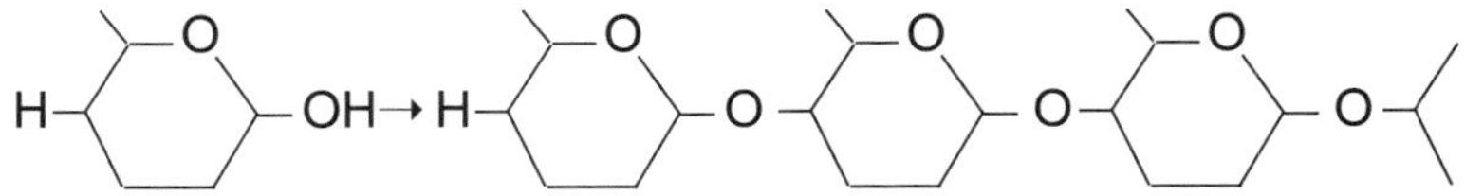

O O O O
H– –OH → H– –O– –O– –O–

Proteins Amino acids link to make up proteins

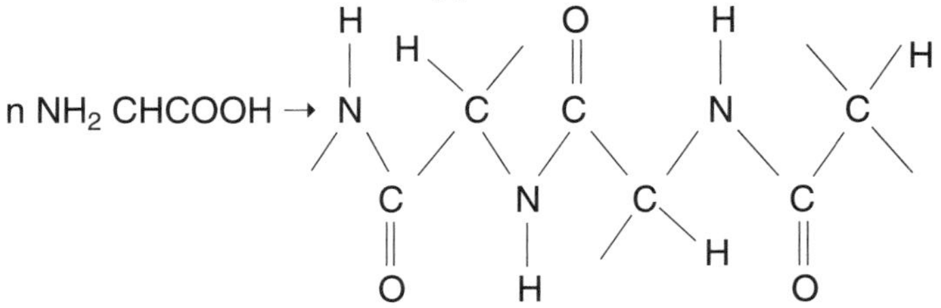

H O H
H H
n NH_2 CHCOOH → N C C N C
C N C C
H
O H O

Synthetic polymers
Many different types exist.

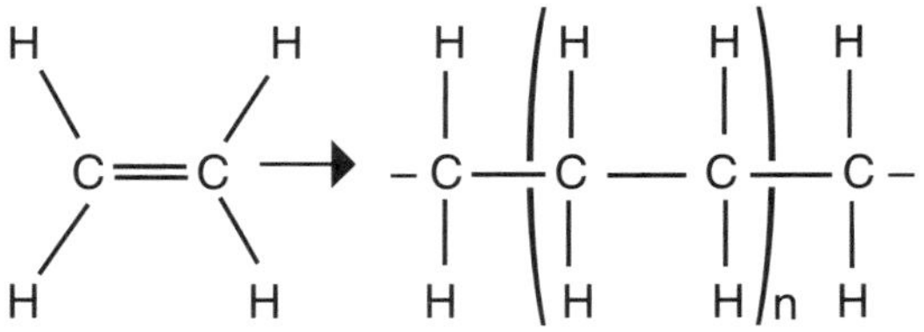

H H H H H H
C=C → –C–C—C–C–
H H H H $H)_n$ H

polyethene

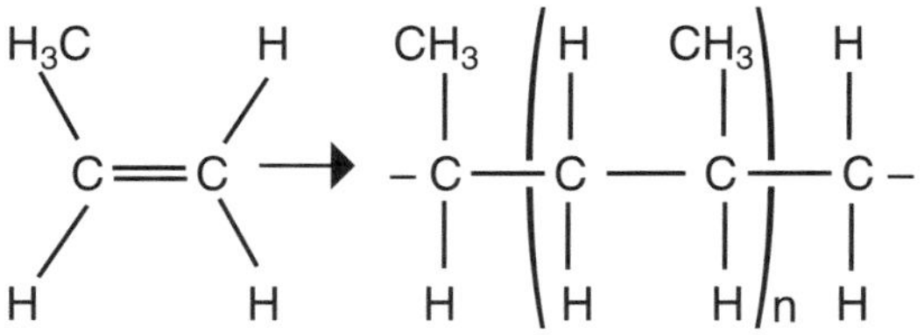

H_3C H CH_3 H CH_3 H
C=C → –C–C—C–C–
H H H H $H)_n$ H

polypropene

Cl H H Cl H Cl
C=C → –C–C—C–C–
H H H H $H)_n$ H

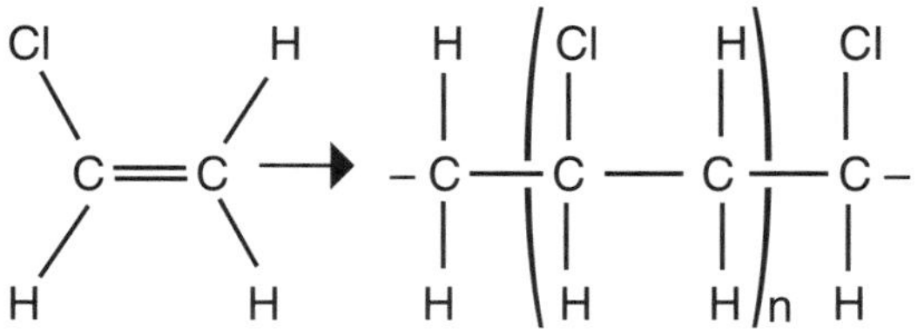

polyvinylchloride